LETTS HOME DECORATOR

STORAGE

LETTS HOME DECORATOR

STORAGE

JILL BLAKE

First published in 1994
by Charles Letts & Co Ltd
An imprint of
New Holland (Publishers) Ltd
37 Connaught Street
London W2 2AZ

Designed and edited by
Anness Publishing Limited
Boundary Row Studios
1 Boundary Row
London SE1 8HP

ISBN 1 85238 378 X

A CIP catalogue record for this book is
available from the British Library

'Letts' is a registered trademark of
Charles Letts & Co Ltd

Editorial Director: Joanna Lorenz
Project Editor: Lindsay Porter
Design: Millions Design; Sarah Kidd
Photographer: John Freeman
Illustrator: King & King Associates
Contributors: Mike Trier, Marion Elliot

Printed and bound in Spain

PUBLISHER'S NOTE
The authors and publisher have made every effort to ensure
that all instructions contained within this book are accurate
and safe, and cannot accept liability for any resulting injury,
damage or loss to persons or property however it may arise.
If in any doubt as to the correct procedure to follow for any
home improvements task, seek professional advice.

CONTENTS

INTRODUCTION

We all collect clutter, and as a result every home needs adequate storage if it is to appear tidy. But in many apartments and houses, there is not much space in which to install cupboards, drawers and shelves, which can lead to problems. This book looks at ways to accommodate all your requirements in terms of your budget and the space available. If you are considering shelving, there are instructions on installation and choice of shelving materials and fixtures. If freestanding units are more appropriate, there is advice on the options available, including buying secondhand, and how to renovate old pieces. There are also suggestions for making the most of otherwise unexploited spaces, and ideas for stylish and inexpensive 'instant' storage.

ASSESSING THE SITUATION

Try to look at your home with fresh eyes (difficult if you have lived in it for a long time – easier if you have just moved, or are actually in the process of moving). Don't rush out to buy something when the clutter reaches overwhelming proportions. Try to make provision for storage at the 'structural stage', especially if it is to be built-in and decorated to blend in with the room scheme. And plan your storage from the inside out so it really will hold all the items you need to fit in.

There may be some 'dead space' which you have not thought about, which can be filled with storage. For example, there is often nothing under a window – except for a radiator, which might be moved. A built-in window seat with lift-up top can provide ideal toy, luggage or sports equipment storage.

Bookshelves can also be practical in this situation, tailored into the angles of a bay. A small flat window may be a good place to site a desk or dressing table with wall-mounted shelves above, placed to each side of the window. The curtain treatment can be designed not to obscure the shelves. This way the room retains most of its floor space and the overall effect is sleek and streamlined, as well as tidy.

OVERHEAD ... AND UNDERFOOT

Then there is the space above your head. In a bedroom or bathroom, storage units can be screwed firmly to the ceiling, combined with tall cupboards to each side, and used to create an alcove – or combined with curtains for a four-poster effect. In the kitchen and utility area, ceiling racks and clothes airers (which are raised and lowered on a pulley) can be practical – as well as decorative, if

they are used to display colourful baskets, drying flowers and herbs, *batterie de cuisine* and shiny copper pans.

But these racks can also be used in garages, for tool/garden equipment storage; in bathrooms, for clean folded towels and bulk-bought lavatory paper; in teenage rooms for sports equipment or accessories; in children's rooms for toys – but consider the safety factor with young children.

The space over the top of the door can be a good place to put a shelf or small cupboard (again firm fixing is essential). And space above cupboards, wardrobes or kitchen units can be filled with shelves, or as storage for infrequently used items.

The picture rail in a tall room can be extended to form a Delft or plate rack (a wide shelf or decorative brackets) which can be used to display a collection of plates, jugs etc, but in the dining-room and kitchen can have a more practical storage application.

The loft or attic has long been the

Above: Storage can be decorative as well as practical; shelves are given an individual touch with a design stencilled in stain on pale-coloured wood. The arrangement of books, ornaments, fruit and flowers all add to the atmosphere in this country-style room. Illuminate such shelving for extra impact.

Left: *A shelf-filled wall makes an eye-catching feature in a square room, creating a library atmosphere – the whole is well lit from above so the book titles can be read, and the desk area used with maximum efficiency.*

traditional place for 'dead' storage – but it might be worth while installing a roof light to provide natural daylight, and adequate electric light, plus a good-quality fold-down loft ladder, and putting some industrial shelving into the roof space. It will probably also be necessary to put floorboards down to create a walkway to the shelving, and to avoid putting your foot through the ceiling.

And what about under your feet? It may be possible to cut a trapdoor in a wooden floor and use the area underneath for security storage – several companies make small safes which are intended to be used in this position. In some rooms, storage can be built on top of the floor to create a split level area – or a conversation pit. In children's and teenagers' rooms, the bed or bunks may well be raised to a practical height, and the space underneath used for wardrobes; or a desk, shelves and chest-of-drawers; or a bench and workout equipment.

There is often a lot of wasted space under beds. If they are on legs, it is easy to install drawer-type storage on castors, making it easy to manoeuvre in and out. Beds may come with their own built-in storage drawers – consider this when purchasing new ones.

Don't forget in the bathroom there is often wasted space under the bath. Bath panels can be purchased which include some storage or shelves, but you can install your own under-bath storage system, and then put openable doors onto a frame to close it off, in place of the traditional bath panel. And an area under a big basin in the bathroom or bedroom can be fitted out with shelves or drawers – in some cases concealed by a pretty fabric curtain to co-ordinate with the furnishing scheme.

REWARDING RE-THINK

There are other areas in the home which can be vastly improved and used in a much more practical way – the traditional under-stair cupboard for example, is often a general dumping-ground. Opening out this area to create a proper 'study' corner with desk, shelves etc, is one option. Or it may be large enough to build in glass, bottle, and china storage; or to create a walk-in cupboard; or to make a downstairs shower room and cloakroom combined.

In a fairly large bedroom, the space may be divided by free-standing wardrobes at right angles to the main walls, backed by other cupboards or chests of drawers to create a separate dressing area and make twice as much storage space.

Above: *Simple storage can often be the most effective – basic white laminated shelves are suspended on metal supports, which are painted to match the recycled trolley and chair frames, to create an integral scheme in a small 'home office'.*

PLANNING AND MEASURING UP

 In a perfect world, the ideal home would have 'a place for everything – and everything in its place'. But as a family grows, and new interests and hobbies flourish, we all collect essential items and hesitate to throw out anything re-usable. As a result, our homes become more crowded with possessions and adequate storage is essential to keep everything clean and safe as well as creating a pleasant home environment.

As a general rule you can never have too much storage – although your possessions will tend to grow to fit all the available space. You may wish to have a flexible system for all the essentials, but may find in many apartments and houses there is not much room in which to install shelves, cupboards and drawers in the first place. Or you may be restricted by a tight budget – and have to use some temporary storage initially until you can afford something more permanent.

SIZING UP THE SITUATION

Try to plan your major storage first. Decide how much is to be provided by freestanding furniture such as wardrobes, cabinets, cupboards, specialist pieces, chests of drawers, under-bed storage drawers, bookcases, and storage chests, and how much is to be provided by built-in furniture. This can be made by a carpenter or joiner to your brief; designed by a company specializing in individual custom-made items; supplied by one of the nationally known bedroom, kitchen and bathroom fitted-furniture experts; or designed and built by you as a do-it-yourself project. The last option need not be as difficult as you think.

Before you start to shop around for freestanding pieces, or call in an expert, it helps to do your homework first, and to have some idea of what you want and what you need to store. For example, in the bedroom or dressing area, make arrangements to store personal items such as clothes, luggage, linen, and cosmetics. In the bathroom, you will need space for medicines and first-aid equipment, but also may wish to store clean towels, cosmetics, spare lavatory paper and other supplies. You may even want to keep some sports equipment in this room if you use it for working-out.

Any linen cupboard needs slatted shelves to allow warm air to circulate to air the clean clothes, but it is not a good idea to leave linens in the warm for too long, so plan dry

Left: This pull-up door is fitted onto shelving to hide clutter. It could also be used to hide a microwave from view.

Right: *The insides of wardrobes have to be as well planned as the room itself, to ensure everything will fit in. Here everything has been measured to make sure – pull-out drawers, shelves and space for shoes and luggage have all been carefully calculated.*

Below: *A refrigerator, however large, is not suitable for storing all the food you need. This freestanding larder incorporates a cold slab for meats; wire baskets for vegetables and fruit; wine racks and shelves for tins and preserves.*

storage for sheets, towels, and table linens elsewhere in the house. There may be space for a cupboard or suitable chest on the landing, or in one of the bedrooms.

In the kitchen, plan space for food, dry goods, cleaning materials, baskets, picnic hampers, glassware, crockery and cooking pots and utensils. If you don't have a separate utility area, you will also need to accommodate washing and ironing equipment, flower-arranging items, and possibly pots and pans for preserving. In the dining area you will want to store cutlery, china, table linen, bottles and glasses; in the living-room allow for books, records, video tapes, cassettes, CDs; in the home office or study – even the garage – you may need to store books, files, papers, materials for hobbies, and do-it-yourself equipment.

In order to plan your storage properly you will have to go back to the steel measuring tape (remember fabric tapes stretch in use so are not reliable), and drawing board to prepare a scale plan, and use this to help you with fitting in the various cupboards and shelves as described in Practical Planning, *see* pages 12–13. It is much easier to plan on paper first, than trying to manhandle large pieces of wood, doors, and supports into a room, and then find they don't fit.

EFFICIENT KITCHEN STORAGE

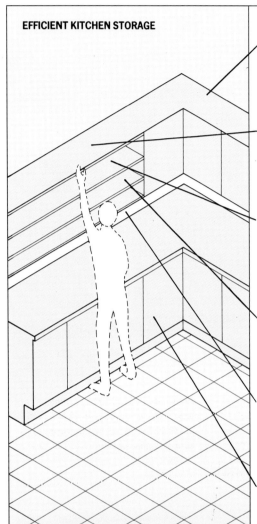

Wall cupboards over a work surface should be a maximum of 30 cm (12 in) deep, to prevent bumping your head as you work.

A top shelf 105 cm (3 ft 6 in) from the work surface will just be in reach for most adults. Heavy objects should not be placed at this height.

Items that are used regularly should be stored about 90 cm (3 ft) from the work surface, so they can be reached comfortably.

A shelf at eye level is useful for cooking ingredients, spices and so on. This may range from 150–170 cm (5 ft–5 ft 8 in) from the floor.

Allow at least 45 cm (18 in) between the lowest shelf and the work surface, to enable small appliances to fit underneath the shelf.

Most dishwashers will fit comfortably underneath a work surface 60 cm (2 ft) deep.

You will also need to think three-dimensionally – and this may involve you in drawing wall elevations as well as floor plans. Height will also be important, and when you are measuring up don't forget the depth and width of details like skirting (base) boards, door architraves and window frames, as well as recesses. These should be checked at several different levels, since walls are rarely perfectly straight.

All these sizes can be vitally important and, if measured incorrectly, can result in total failure, as the shelving and supports, hanging rails or pull-out baskets and drawers simply won't slip into the recess as you had hoped – or the window seat with lift-up top will not open without banging against the sill. This means accurate arithmetic is essential, so if you are measuring for doors and shelves, make sure you get the true vertical (for struts or slotted-angles to support the shelves) by using a plumb line, and the true horizontal by using a spirit level, and always double-check your figures. It is worth getting somebody else to take the measurements again as a precaution, and always insist any carpenter, joiner or built-in furniture company do their own measuring up, so you can't be held liable for any mistakes.

It is also worth mentioning, when you are discussing this type of custom-made furniture and obtaining estimates (for time as well as cost) from the experts, that you should check exactly what you will get for your money. Some companies produce a completely finished product down to the last handle, while others do the carpentry and joinery and leave you to finish it off, add beading and handles, and decorate it yourself.

Left: *The garage can be used to store tools and as a workroom. Storage facilities can be fairly basic here. Measure up and make sure the height, depth and strength of the shelves are adequate and will take everything you want to store neatly and tidily.*

NEVER ENOUGH SPACE

We all want cupboards which are larger on the inside than they appear from the outside, especially if the room is small. If you do have to install a bulky cupboard in a small area, mirrored doors can help to create an illusion of space. There are some special track systems which allow you to hang such doors on a track to fit across recesses, or to create a complete storage wall. In an ideal world storage should be planned from the inside out. Always measure up exactly what you want to store, and relate these figures to the width, depth and height of drawers, shelves and cupboards. Take these sizes with you – and your measuring tape – when you go to buy items of free-standing furniture, or to discuss a project with a carpenter or fitted-furniture expert.

In the dining area for example, measure all the dimensions of stacks of plates, bottles, glasses and serving dishes. In the living-room, check on the size of books, record sleeves and tapes. In the bedroom, measure the length of your longest dress, width and length of winter overcoats (on a hanger), and width, depth and height of luggage. Again, plan ahead and think about future storage possibilities. In some cases it makes sense to have adjustable shelving, pull-out racks and drawers inside freestanding or fitted furniture, rather than fixed shelves, so you don't outgrow your storage.

Also remember to plan lighting to relate to the storage. Accent lighting can be incorporated into display shelves for maximum effect; bookshelves and office-type storage should be well lit so you can see titles and other information at a glance; and bedroom and bathroom cosmetic storage might include an illuminated mirror. Lighting deep and walk-in cupboards is always worthwhile – fit these with door jamb controls so the light comes on automatically when you open the door.

Above: *Building your own cupboards? Here doors have been fixed across a recess, and the interior fitted out with plastic stacking crates for additional clothes storage. These are inexpensive to buy, and you can add to them as your needs increase. Look in* kitchen and household departments for ideas to see what you can adapt to suit your purpose. Always measure the inside of the unit or cupboard before you buy interior fittings.

Right: *Plastic stacking crates can also be fitted into kitchen units and used in utility and laundry areas – again measure the interiors of the units accurately.*

PRACTICAL PLANNING

The best way of working out how to fit in items of freestanding furniture and built-in storage is to 'get it right' on paper first, by making floor plans to scale of each room. Plans of walls are called elevations, and you may need to make these too when it comes to working out positions for your storage – to make sure heights are correct; pieces will fit under windowsills; shelves can be reached easily, but are not so low as to present a safety hazard, etc.

It also makes sense to do an overall floor plan to show how the various rooms relate to each other – and to the hall, stair and landing area. This master plan will enable you to cope with any structural alterations at the outset – installation of plumbing pipes and cables, cutting holes in walls and ceilings to accommodate electrical fittings, and installing built-ins are all classed as structural jobs.

ACCURACY AT ALL TIMES

Measure up accurately using a steel rule or tape (fabric ones stretch in use), allowing for projections and recesses, and existing built-in items. It helps to have a 'mate' to hold the end of the tape and to double-check your arithmetic. Make sure you measure on the true vertical and horizontal – use a plumb line or spirit level to check levels.

Plot positions of electric points and socket outlets; plumbing pipes etc. Don't forget to 'think three-dimensionally' and measure depth and height of skirting (base) boards, any dados (chair rails), cornices, covings etc; height of the windowsill from the floor; depth and width of sills, window frames and reveals, architraves etc, since such projections may have to be accommodated when you are installing storage, and will need to be plotted on any elevations.

MAKING THE PLAN

Draw out the plan to scale on either plain or, better, squared paper – you can get this from graphic design shops, or use a child's arithmetic book. Choose the scale to suit yourself, but 1:50, 1:25 and 1:20 are the most usual to use (1 cm to 50, 25, or 20 cm of room space) or, if you are not metricated, 1 foot of room space to ½ or 1 in on the plan. Use a set-square and a ruler to ensure straight lines and accurate angles.

Make templates, in the same scale as the floor plan, of all the items of furniture, bathroom or kitchen equipment you intend to put into the rooms – allow for existing items as well as any proposed purchases. It helps to colour them, so they can be seen on the plan more easily – colour-code if necessary, using one colour for existing pieces, another for proposed items, and a third for 'possibles', and so on.

You can buy special architects' transparent templates, in various scales, from graphic design and artists' supply shops to make this easier – just trace through the shape onto plain card or paper and cut out. Remember to buy the same scale of furniture template as your floor plan.

Below: Planning your room on graph paper ensures you will get it right from the start. Draw out the plan of your room to scale, and cut out furniture to the same scale. Position the furniture on the floor plan until you have achieved a pleasing and practical arrangement.

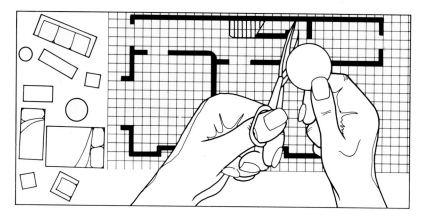

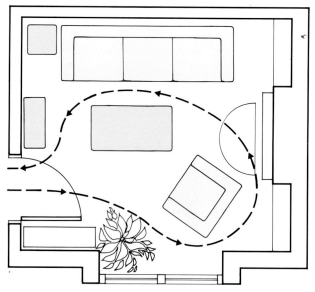

Above: *Don't forget to allow movement around the room. Plan for access to storage facilities and sockets, and room to open doors, drawers and windows.*

illuminate the insides of deep cupboards. Also plan so that water inflow, and soil outflow pipes have the shortest runs to and from service appliances and sanitaryware. Provide power supplies for appliances and equipment, TV, video, computer terminal, music centres etc exactly where required.

THE OTHER DIMENSION

Elevations (flat plans of walls) are made in the same way. You will need to measure carefully and plot the exact positions of doors, windows and fireplaces. Remember to allow for frames and architraves and to measure the depth of skirting (base) boards and mouldings so these can all be shown accurately on the elevation.

You can draw these in relation to your floor plan so you get an 'exploded' impression of the entire room; this could be used eventually to make a model of the room if you want to test the effect in three dimensions.

FITTING IT ALL IN

Once you have completed the paperwork, and established the accurate positions of any freestanding furniture and necessary equipment, you can work out the best place for your built-in storage cupboards, shelves, sets of drawers or whatever.

Recesses to each side of a fireplace are often the natural place to install wardrobes, shelves and drawers, but don't forget you need adequate depth to make hanging

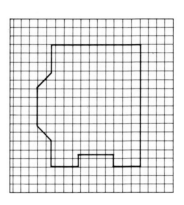 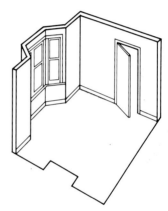

Above: *Use a scale floor plan and accurate measurements as a way of working out the third dimension. This will enable you to see whether storage will fit into recesses, under window sills or other awkward areas.*

Move the cut-out shapes about on the plan until you reach a satisfactory arrangement. Then you can draw in the furniture positions on your final plan, but always work with templates first as it cuts down on time, and prevents you from having to keep drawing and redrawing the plan.

Remember to allow for the opening of doors and windows; pushing chairs back from tables; moving round furniture; crossing the room – this is called the 'traffic flow'. You will also need to consider the opening of doors and drawers on furniture and cupboards, so that they can be used easily. You will need to have accurate measurements of the widths and depths of these. Once the positions for furniture, appliances and equipment are firmly established, you can plan lighting, plumbing and power supplies so they come exactly where they are needed. Lighting can be centrally positioned over a dining table for example, or planned to illuminate a desk, hobby bench or work surfaces in the kitchen clearly, without glare or shadow – or to

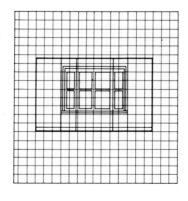 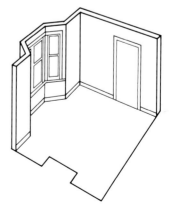

Above: *Work out wall elevations to scale, and plot accurate positions of doors, windows and chimney breasts. You can then see at a glance whether there is room above, or beside a door to fit in shelves and storage.*

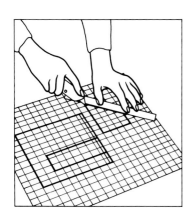

Above: *Paste a scale floor plan onto card; do the same with the wall elevations and cut out carefully with a craft knife, using a cutting mat or wooden board.*

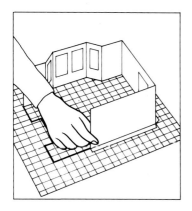

Above: *Bend and fit the cut-out wall elevations accurately, so they relate to the floor plan, and form a three-dimensional model of the room.*

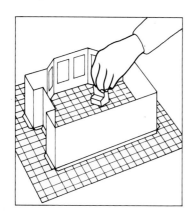

Above: *Make shapes of the furniture to the same scale and move them about inside the model room to see where they will fit best. You can then see where to site your storage.*

clothes or storing most other items a practical proposition. It is often necessary to design storage so that it projects beyond the chimney breast – this can sometimes then be linked·with narrow shelves across the breast.

In a small, square room you can sometimes make the most of corners by installing a corner cupboard across one or more of the right angles. In some rooms, such as a dining room, this can add charm and character by giving the room an unusual shape. But the interiors of such cupboards need very careful planning if there is not to be a lot of wasted space at the back of them.

You will find the floor plan will help you to assess the volume of the interior of any built-in furniture, but don't forget to allow for the opening of the doors or drawers of your planned additions. You can work this out by drawing in the door swings, and the depth of open drawers, with a dotted line on your plan.

The elevations will help you to assess the height of proposed items – you will be able to see whether they fit under windowsills; under or to the side of any wall-mounted light fittings, which may have to be moved; or to the side of doors and windows. You can also work out the suitable position of pull-out racks, hanging rails etc in relation to the size of the items you want to hang and store.

This planning method will also help you to judge how easily you will be able to reach things stored on shelves and in cupboards. Measure your own height and draw a red dotted line on the elevation at your eye level, so you will know whether you will be able to see what is stored without having to stand on a stool. Remember you can stretch your arms about 45–50cm (18–20in) above eye level. If you need to install very tall units you should consider a step stool or set of library steps in your budget costings.

Above: *Bolt-together metal angle shelving fits neatly above the cistern in this bathroom, and combined with shelves makes maximum use of what could have been wasted space.*

The flexible system enables shelf height to be varied to accommodate different items.

SPACE SENSE

Good storage only works if it houses all the items you want to put away neatly and efficiently – this means that clothes don't come out looking crumpled and needing ironing before you can wear them; that you can easily take out much-used objects without having to move several layers first; and there is little risk of breakage or damage to fragile items. As in most aspects of homemaking, in the final analysis it comes down to practical planning and making commonsense decisions.

Don't forget, it is essential to measure all the items you want to store before you start designing any storage system – in this way, if you plan units, chests of drawers, wardrobes, cupboards or shelving 'from the inside out' you will end up with as near a perfect system as possible, to suit your personal requirements. Don't forget weight as well as size – some things (like books, *see* page **44**) can be very heavy, so consider this when working out the supports needed for shelving. Drawer runners and drawer bottoms must also be sturdy, so they don't collapse when filled.

FLEXIBLE APPROACH

Domestic storage systems should be both flexible and versatile – try not to limit them to specific areas, although you will obviously need to store clothes in or close to the dressing area; items for the dining table in the dining room or kitchen; *batterie de cuisine* near the stove and food preparation area; books and files near the desk or computer terminal; records, cassettes, CDs and videotapes alongside (or underneath) the music centre and television set, and so on.

A combined unit not only gives a room a more spacious appearance, but it is more practical where there are a variety of things to store. Choose a modular system which can be combined to fill any width to within about 30cm (12in) at either end of a row of units – these gaps can be filled if necessary by scribing a batten to the wall.

Where possible, choose a system to which you can add supplementary units, extending along the wall or upward to the ceiling. In this way you will be able to accommodate extra storage in the existing system as and when you need it or can afford it. Look for ready-made ranges which will remain in stock for some time (some of the flat-pack or knock-down [KD] ranges tend to be here today and gone tomorrow). If you make your own, or have units custom-built, it should not be too difficult to expand the system.

Flexibility within the system is also important. Adjustable shelves are much better than fixed; pull-out or lift-up drawers and trays are very useful; pull-out rails and pull-down rails on several levels are practical in wardrobes or clothes cupboards, especially those less than 50cm (21in) deep, or which are tall or where the top layer cannot be easily reached.

Open shelves, units and dressers can be an excellent choice for living-room, dining-room and kitchen storage, and at the same time will provide an attractive display which will give character to a room. But valuable books, fragile glass and china and other precious items are often best protected. This does not mean shutting them away with solid doors – they can be kept behind glass to protect them from dust, or behind a decorative wire mesh.

Above: A blank kitchen wall is converted into a dual-purpose dining area and storage space. A flap-down semi-circular table allows for extra circulation.

Above: When it is in use, there is place for two diners at the table seated on space-saving folding chairs. The shelves above can be tailored to hold items needed for the

dining table. The whole unit is neat and slim and takes up very much less space than a conventional kitchen unit or dresser.

LIVING-ROOM STORAGE

This usually includes books, records, magazines, cassettes, videotapes, bottles and glasses. Magazines should be kept flat, or between rigid divisions that prevent them from slipping down; books need to be stored upright so the titles on the spine can be easily read; video and audio tapes and CDs are much the same, and LPs must be stored upright or they will warp. Shelves for all these should have a convenient divider every so often, to enable you to flick through them or remove one without the whole row collapsing.

There are many prefabricated racks and dividers available which you can slot inside cupboards or units, as well as some free-standing and wall-mounted storage systems which are designed specifically to hold these various items. Again you will need to measure what you have (especially with books) for height, depth and width to help you get the sums right – and also to calculate the volume of space required. Think ahead too – you are bound to add to a collection of CDs, tapes, books or whatever.

Drinks, bottles and glasses also need careful storing – many ready-made systems are not tall enough to take an extra-large bottle, or a special decanter; glasses also need adequate facilities to store them upright, singly (glasses stacked inside each other get broken) and away from dust. They are best placed on narrow shelves, one glass deep or slotted (wine glass stems) into a special rack which can be built into the top of a cupboard – sometimes these can be designed to pull out. Table wines should be stored with the bottles lying down, to stop the corks from drying out; other bottles should be stored upright.

Ideally such items need to be stored near a flat surface, on which you can pour, mix and serve, with the bottles at eye level. In some cases a wall-mounted or tall freestanding cupboard with a wide shelf or drop-down flap is the answer; or a base unit or drinks cupboard with flexible, adjustable shelves inside and generous-sized wipe-clean top for pouring drinks. There should be at least one adjacent drawer to hold corkscrews, bottle openers and other bar equipment.

Many of the above comments also apply to equipment for the dining table. Cutlery should be stored in special divided drawers, or in racks or baskets to avoid general jumble and to prevent scratching; china should be stored so that it is easily accessible, and then not too many piles of plates, dishes etc on top of each other – china can be very heavy. Again the storage area can be combined with a flat, wide top for serving, and drawers for mats, napkins etc.

BEDROOM STORAGE

Clothes usually take up the greatest amount of storage space in a bedroom and cause the most concern. The ideal situation is a separate dressing room situated between the bedroom and bathroom, but most of us lack enough space for this, although it may be managed when a bedroom and bathroom are *en suite*, or if a large bedroom can be divided to form two separate areas. Other less frequently used items, such as luggage, or hobby and sports equipment, traditionally kept in the bedroom, need not be stored here – measure up what you need to store and see where you can fit it in elsewhere in the house or apartment.

Most garments are better hung than stored flat, but examine your clothes carefully. Some items, such as sweaters and sports clothing, are better folded or rolled and stored in drawers or on shelves. Fold or roll these and measure them folded – the depth, width and height of the resulting 'parcel' will help you to calculate how much space you need.

One point is worth noting when planning wardrobes and cupboards – clothes are an excellent noise insulator. So if you have noisy neighbours or loud children, you can fit a run of cupboards along party or dividing walls. If the cupboards have backs, fit roof-insulating material down behind (between the wall and the cupboard back) to reduce noise nuisance still further.

Cosmetics can be a problem to store in either the bedroom or the bathroom, because of the enormous number of different shapes and sizes of containers, boxes and bottles. Small items can be fitted into a tray or specially divided drawer; others could go into wire racks or narrow shelves fitted behind cupboard doors (bedroom built-in specialists have designed some ingenious solutions); larger items can be kept in a deep bin with a lift-up or sliding lid. Again measure to work out maximum and minimum sizes.

LINEN STORAGE

Bedlinen, spare duvets, pillows, blankets, towels, tablecloths, napkins, tea towels etc all need to be stored flat and where they will

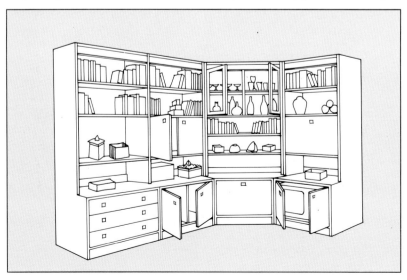

Left: A freestanding shelving unit in the living-room makes use of a corner space, and includes a range of storage facilities. Glassware and drinks are on display behind glass-fronted doors; the television can be hidden away when not in use; and cupboard dividers keep records upright.

Right: *Pull-out wire storage baskets are sold as unit interiors; they can be adapted to other purposes. Here they are fitted under a long laminated worktop which doubles as a dressing table, and are used to store clothes, linen, cosmetics etc in a dressing area. The collapsible metal folding dress rail is another practical item, providing flexible, transportable, hanging clothes storage.*

be kept crisp and clean. Although they may be aired after washing and ironing in an airing cupboard with slatted shelves (usually situated next to the boiler or hot-water tank), it is not wise to keep them in a warm atmosphere permanently because they can rot and discolour.

It makes sense to store bedding as close to the bedrooms as possible, so a proper linen cupboard with wide shelves on the landing, or in one of the bedrooms, is a practical solution. A chest of drawers or cupboard with adjustable shelves can be a good place to store folded towels and smaller linens; an old linen chest or ottoman with lift-up top can take bulky bedding. Items used in the dining room or kitchen might be stored in a special unit in either room or in the hall if it is large enough to take a cupboard, chest of drawers, linen chest or cabinet.

Again items should be folded, and then measured to work out space requirements. Double sheets and duvets take up more space than the single size, and proper bath sheets, as opposed to smaller bath towels, can be quite difficult to store, especially if they are very fluffy.

Dirty linen is usually dropped into a linen bin or basket kept in the bathroom or bedroom, but some fitted bedroom furniture contains built-in pull-out linen bins. You could copy this idea with plastic bins or crates.

TOYS AND GAMES

Children's games and toys need to be simply stored in a system which the children can manage themselves, in the hope that they will be tidy and put things away after use. Narrow cupboards with sliding doors; plastic

stacking crates; built-in storage under beds and bunks; and wheeled boxes with lift-up tops are all options.

Books and board games need more tailored storage of the type mentioned under living-rooms; train and racing-car layouts are best fixed to a permanent board which might be hinged, to flap up or down when not in use; these can be drawn up by a pulley to the ceiling to clear the floor when needed.

Very large toys cannot always be stored successfully, and may have to be stacked under beds or left out, but smaller items should be measured and the alloted space worked out as with all other items. Similar considerations apply to hobby, sports and games equipment. It often makes sense to store outdoor games equipment as near to the front door as possible, and indoor hobby things near the area designated for hobbies.

DESIGN DECISIONS

Each room in the home, and each family, will have individual requirements which will evolve and change over the years. In some rooms it will be sensible to plan ahead for future acquisitions, and possible changes in lifestyle. With children's rooms, in particular, it is important to try to plan a room which 'grows with the child' as they so quickly outgrow scaled-down furniture and nursery-rhyme decorations. But the need for more storage in bedrooms, living-rooms, and kitchens can also increase as you take up new interests and hobbies, buy more clothes, and add to your gadgets, appliances and kitchen equipment.

FITTED, FLEXIBLE OR FREE-STANDING

There are different types of storage units and shelving which you can choose to accommodate your possessions, and most houses end up with a combination of all four.

Fitted cupboards and wardrobes will maximize the storage space, and the interiors can be planned to suit your individual requirements (*see* pages 12–17 on measuring and space required). They are most often used in kitchens and bedrooms and may be designed to fit into chimney-breast alcoves; to create an entire storage wall; to fill an awkward corner; to form an alcove for a bedhead or dressing table; and to create a streamlined kitchen or study corner. They can be custom-built by a carpenter or specialist companies, or made at home.

Above: *A full-height larder is designed with maximum efficiency – the ample door backs include storage racks, almost doubling the space available in the unit. Large cooking items are stored on wooden wall racks.*

Left: *A ceiling-high storage shelf in a beamed kitchen is used to display and store a collection of pots, cups, saucers, plates and bowls.*

Above: *In a small bathroom, a wall-mounted cupboard has been designed to take essential items. The painted façade makes it an interesting feature in its own right.*

Right: *Linen storage does not always have to be confined to the bathroom – a freestanding unit can go into the utility room or stand on the landing. Drawers can be used to store irons, cleaning materials, and hair dryers.*

Modular units are built up from shelves, drawers and cupboards, and come in a variety of different shapes, styles and sizes (modules). They can be fixed or freestanding and so present a flexible way of providing storage, as they can be altered and adapted and added to at will. They can also be dismantled and taken away if you move, or put into another room in a different arrangement.

Various modules can be placed on top of or beside each other, and you can also use them as room dividers (with the weight at the bottom of any tall units so they don't topple over). These can be ready-made, so all you

need to do is place them on top of each other and fix them to keep them steady. Others may have to be fixed to a wall, and some types come in kit form.

Freestanding drawer units and cupboards can be ready-made, or available in kit form. Drawers can be an integral part of a piece of furniture (chest of drawers, desk, kitchen unit, drawers within a larger freestanding item) or incorporated into it by using a kit, or some form of wire rack or pull-out system. Ready-made drawers can also be used under beds. (*See* pages 70–71 for under-bed storage ideas.)

Flexible storage systems usually consist of shelving, which is fixed to the wall by means of slotted metal uprights with brackets which fit into them to hold the shelves, and can incorporate doors and drawers. Some industrial shelving is also flexible, since it can be unbolted and re-made in a different form (*see* pages 32–45 for shelving ideas).

If you buy a ready-made system which has to be assembled from a kit, make sure you have all the parts, fittings, fixtures and screws before you start building. Also, check that you understand the instructions, and that those supplied with the kit are intended for the item you are making.

Having sized up the situation in your home
and considered the type of storage you will
buy and fit (or have fitted), you will need to
think about style, as obviously you will want
your storage to fit in with the basic theme of
the various rooms – and at the same time
possibly to provide some textural contrast.

Items of freestanding storage furniture can
be chosen, like the rest of the furniture, to
create a certain period flavour or a specific
modern look, but built-in and fitted furniture
and shelves are less easily fitted into a style
category, except possibly in the bedroom,
where different woods, finishes and looks are
all available.

You may decide that the wood or other
material you choose for storage and shelving
will help set the scene. Pine shelving and
cupboards would be suitable for a country-
style living-room, bedroom or kitchen for
example, or mahogany for dining-area storage
or in a bedroom where a Regency or heavy
Victorian style is required. Black ash or
charcoal-painted cabinets and shelves might
look good in a city penthouse; white
laminate, or brightly coloured storage in a
modern setting; and sleek steely metal and
glass in a hi-tech environment.

If none of these is appropriate – or
affordable – you may well decide to paint the
storage to blend in, or to contrast, with your
room scheme. You could perhaps use a
painted technique such as dragging, marbling,
stippling, sponging, or stencilling. In an ideal
world, the main decorating in a room should
be done first, except for the final finishing
coat on woodwork, and then any fitted
furniture installed when it has been primed
and undercoated, and the finishing coat and
final touches should be added at the very end.
If you are employing a professional to build
the system for you, discuss the timing so that
the units arrive when the decorating is at the
appropriate stage.

LIVING-ROOMS

Your storage requirements for this room will
depend to an extent upon your lifestyle, but
most of us want to store books, records,
cassettes, CDs, tapes, and possibly bottles
and glasses. In dual-purpose living-rooms,
you may also want to store items for the
dining-table, hobbies, studying, even the
children's toys.

Try to make the most of any natural
recesses, at each side of a fireplace for

*Below: A freestanding wooden storage unit
is used for display purposes. It is coloured to
co-ordinate with the wall.*

example, where you could build in low units to take records and tapes on one side, with the television set, video, music centre, and CD player placed on top. On the other side, a similar unit or cupboard could hold bottles and glasses, or be designed as a desk. Shelves above these units can be used for books, files, or display purposes.

A bay window often forms a natural recess, yet is frequently overlooked. It is an ideal place in which to build a window seat, with lift-up top which can be used to clear away children's clutter at the end of the day, or to provide long-term storage. It is also a good place to site low shelving for books or records. Don't forget you have to allow for this when planning any window treatments. Sometimes a freestanding ottoman or storage box – even a wicker theatrical hamper – can be used for a quick clear-up and can stand in the bay of the window.

If your room is square and box-like, you could make a feature wall out of fitted or freestanding furniture. This can fill an entire wall, and you might have space left in the centre for an electric or gas fire (freestanding or fixed, a gas one working on a balanced flue). On other flat walls you could install shelving, or use a variety of freestanding pieces, such as chests, sideboards, and cabinets, to add a decorative touch.

It is sometimes possible to install shelves for books above a door, and if you are making an open-plan room from two smaller ones and plan to close up one of the doors, this could be made into display and storage shelves. Leave the architrave in place as a decorative feature on the room side; remove it on the hall side and fix plasterboard or hardboard flush with the hall wall. Build shelves across the architrave on the living-room side, and remember to light them interestingly.

Above: *A built-in corner unit for television, video and music centre is flanked by storage cupboards – the result of an economical mix of ready-made and do-it-yourself carpentry.*

The painted finish links with the picture rail, creating an integral designer feel in this modern living-room.

Above and left: *Now you see it, now you don't – a clever way of hiding the television set from view. Purpose-built storage provides a focal point in a room without a fireplace;*

lower units to the left and right house bottles, glasses and records, and the central unit provides storage for the television set.

DINING AREAS

Whether the dining-room is a separate room, part of the living area, or sited in one end of the kitchen, the main items you will want to store are those which you will need for the dining-table – dinner services, glasses, mats, table linen, and candles. You may also want to store bottles of wine, and bottles and glasses for aperitifs and liqueurs. Measure all these items in order to plan your storage as efficiently as possible – remember items like large serving dishes are long and wide, and some bottles are unusually tall.

You will also need a serving surface – somewhere to place serving dishes before they are taken to the table; to set out food for a cold course; to stand plates to warm; or to keep food hot. Sometimes a trolley (the warm-shelf, 'hostess' type possibly) is the practical answer, especially as it can be wheeled back and forth between the kitchen and dining-room. But units, freestanding cupboards, cabinets and sideboards can often provide storage space combined with a serving top. Make sure any such surface is adequately illuminated.

In dual-purpose dining-rooms you may also need to provide desk or study facilities, so if you have two recesses to each side of the chimney breast you might use some of the ideas suggested for living-rooms with dining-table storage on one side – or again you could incorporate all this into a complete storage wall. Again, think three-dimensionally and look above your head – a Delft rack, usually positioned above the picture rail, running around the entire perimeter of the room can be used to display plates or other objects, but this can also be used to store them. This creates a traditional look and can also be carried through into the hall or kitchen, if it suits the style of the property.

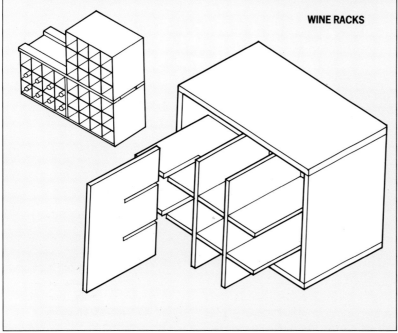

WINE RACKS

Above left: *Glass-fronted cupboards are both practical and decorative, and house glass and china in a dust-free environment.*

Left: *Wine racks are preferable to boxes for keeping wine in top condition. A crate can be divided with plywood partitions slotted together with halving-joints.*

HALLS

This area of the home is often very small and narrow, but is somewhere where you need to be able at least to hang visitors' coats, put the telephone, keep umbrellas and walking sticks, store the post and keys, and check your appearance before leaving or opening the front door.

The way you provide such storage facilities will depend on the space you have and the style of your home. In narrow spaces, some wall-mounted shelves and a wall-hung mirror may be all you have room for. If there is a bit more space, use a hat stand, or even a refurbished old-fashioned hall stand, and a large pot or umbrella stand for wet umbrellas and walking sticks. In a larger hall you can provide storage with freestanding units; a chest of drawers; a hall table; or a built-in desk and shelves.

One area in a hall which is often overlooked is the space under the stairs. This is often just left to accumulate clutter, but it is sometimes possible to open out the area under the stairs and fill it with purpose-built storage, which might incorporate a wine rack, space for files and papers, a telephone and a desk top. If you need the under-stair space to

house items like the vacuum cleaner, outdoor coats, and gas and electricity meters, these need to be hidden from view, but it may be possible to install shelves, wine racks, wire pull-out drawers, and stacking crates to keep everything tidy and under control. Don't forget you will again have to measure everything you want to store to make sure it will all fit in place neatly.

STUDY OR HOME OFFICE

The storage required here will depend on the type of equipment you need. If it is merely to store a few books, papers and pens, or is a corner where you do the accounts, or the children do their homework, you will probably be able to manage with a flat-topped desk, or a bureau with drop-down flap. A base unit, or wall-mounted drop-down shelf could also be adapted for the purpose, to form a writing surface.

However, don't forget to think the project through and plan for any hobbies as well. For example, if a work surface is that bit larger, or stronger, it could be used to hold a sewing or knitting machine. Papers and pens can be stored in the desk drawers, and a few wall-mounted shelves or a bookcase can be

used for files and reference books. Don't forget to plan adequate task lighting for such areas.

If you work from home, and you have more complicated requirements – including needing to store drawing materials; make provision for a computer, VDU, or drawing board; store tapes, discs, reference books, or samples, as well as perhaps directories and reference books – you will need flexible shelving and storage arrangements which must be strong. Some of the systems described in Focus on Shelving (*see* pages 32–45) may well be suitable. Alternatively you could use a fixed system, and close it off from view with blinds.

Above: *A home office system to put together yourself relies on a series of modules to create desk, storage, shelving and filing facilities.*

Left: *A pull-out shelf is designed to store the keyboard of a computer in this home office tucked into one corner of a bedroom.*

KITCHENS

The bulk of kitchen storage is provided by base units, cabinets and tall or wall-mounted cupboards, which become an integral part of a built-in kitchen – lower units are usually topped by a work surface. Space can be left between cabinets for appliances such as dishwashers and washing machines. In large kitchens, base cabinets can also be used to form a peninsular unit, or to make a natural division between kitchen and dining area.

Right: A large walk-in larder unit provides more than adequate storage for bottles, dry goods, herbs and spices, and is neatly slotted into a corner.

Below: Well-planned storage within a kitchen unit – this deep drawer is fitted with a wire vegetable rack.

Often racks for wine storage, or a roll-around trolley or portable butcher's block can be fitted under such units.

Once again you will need to work out exactly what you want to store, and do some measuring up to make sure your preferred units will provide adequate storage space. There are many interior fittings available for specific storage to expand the use of such cabinets – to make the most of a 'blind' corner space; to hold tins or vegetables, an iron and folding ironing board; keep appliances such as food processors off the worktop – with varying degrees of sophistication from built-in chopping boards, pull-out pantries, wire racks and drawers, to revolving carousels. Some of these can be provided by the unit manufacturers; others can be bought as separate units.

Rigid (and self-assembly units) come in various modular sizes which are not always practical to fit your room and to store your possessions or, if they are the correct size, they may not be in the right style for your kitchen. If this is the case, call in a purpose-built kitchen specialist – or make your own unit – and plan from the inside out.

Currently there is a trend towards the 'unfitted' kitchen, with various items like a butcher's block on wheels, larder cupboards, and dressers providing much more flexible and movable storage with a casual country-style image. If you can't afford the more expensive ranges, you might well adapt some of the second-hand finds, which you can refurbish (*see* pages 51–57), creating a continuity of colour and pattern with paint, stain or self-adhesive covering, which will help you to link several pieces.

MAXIMIZE EXISTING SPACE

It is often possible to create more space in existing units and cupboards if you use your imagination. The backs of cupboard doors, for example, can be fitted with various shelves or racks for small items, or with specialized fittings for tin foil, kitchen roll (paper towels) and cling film (plastic wrap); shelves and drawers can be fitted with special plastic-coated wire organizers to take saucepans, pan lids, china or table linens; other racks are specially designed to take china, plates, glasses and mugs. The insides of drawers can be fitted with special segmented insets to keep cleaning materials or bottles upright.

There are many ready-made wine racks available; some are collapsible, and some are triangular (ideal for under-stair cupboards); some take individual bottles (and these need not be wine) in each pigeon-hole, and others take larger ones for bulk storage. These can be fitted into units, but you can also make your own by gluing together lengths of PVC piping of suitable diameter cut to suitable length to fit your cabinet or cupboard.

The backs of larger cupboard doors, or the sides and backs of long cupboards can be organized to take brooms, mops, irons and

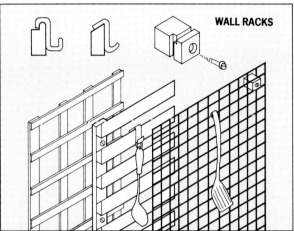

WALL RACKS

Above: *Ready-to-hand cooking implements can be stored in a number of ways, including being suspended from a specially designed wooden rack.*

Above right: *Wall racks will keep kitchen utensils close to hand, and can be a decorative feature of the kitchen. A wire trellis can be fixed to the wall with wooden blocks. Saw half way through the blocks, creating a groove to hold the trellis, and screw to the wall. The utensils are suspended on S-hooks. Wood can be used to make a trellis, or to fit rails across a framework. To hang the utensils, bend metal hooks to fit over the rails.*

Right: *Cooking implements are suspended from a metal bar above this stylish gas cooker in a modern setting.*

ironing board, or the spare hose to the vacuum cleaner – such things are always better hung up. There are many simple notched, grooved and hooked gadgets which are specifically designed for the purpose, either made of metal or plastic-coated wire, or you can tailor-make your own from notched wood or dowelling.

Existing shelves, or the underneath of wall-mounted cupboards can have speciality racks fitted, which hold glasses by their bases or stems, spice jars by their lids or knives in a special safety block. An even simpler device is to screw cuphooks into the edge of a shelf, or under the bottom of the cupboard on which to hang cups, mugs, jugs, or a variety of cooking implements.

Pots and pans and other cooking equipment can be hung up tidily in close proximity to the cooker or work surface. There are special hanging pan racks which can be suspended from the ceiling above a worktop, or an adjustable ceiling airer on a pulley can hold baskets, drying flowers, even linen. Special steel bars with clip-on hooks; magnetic rods; shelves with movable hooks; wall-mounted wire grids; special racks with wooden slats; and pegboards with holes and pegs on which to suspend handles are all good methods to use. Where space is short, a heavy-duty metal chain slung from two ceiling- or wall-mounted hooks can be a flexible alternative. But remember most of the items you will want to suspend can be very

heavy, so it is essential that any racks, hooks, chains or other means of support are firmly fixed to the ceiling or wall.

In the kitchen, as with other areas in the home, you can make a decoration out of a necessity. A kitchen dresser can store – and display – china, canisters and storage jars. Prettily shaped transparent jars can be used for pastas, pulses, dried fruits and herbs. Attractive storage caddies can be used for flour, sugar, teas, coffees, and biscuits, and there are many different types of spice storage racks, mug 'trees' and handy baskets for cutlery, and bottles which are decorative in their own right. A simple, basic crock can be used to hold wooden spoons, cooking forks, spatulas and other equipment.

BATHROOM AND LINEN STORAGE

Some of the kitchen storage ideas can be adapted to the bathroom. For example, children's bathtime toys, sponges and loofahs can be suspended by some of the methods suggested for implements in the kitchen. Many of the kitchen-furniture manufacturers also make bathroom units, but you will discover they often have not thought about the practicalities of storing lavatory paper in bulk, the different heights and shapes of shampoo bottles, and the different sizes of folded towels. Once again it is a question of sitting down and working out exactly what you want to store in the bathroom (often you are limited by space), measuring up, doing some sums and working from the inside out.

There are various ways of making storage space in a bathroom. The panel at the side of the bath, for example, can be converted into storage, and some ready-made panels are already adapted. If you have enough space, tall cupboards placed at both ends of the bath, linked by top cupboards above, will create a cosy bath in an alcove, with somewhere to fix shower curtains to draw across totally enclosing the bath. There may be space to install wall-mounted shelves at the side of the bath (position these high enough up to avoid splashing); to each side of the window below; above and to the side of the basin or in any alcove. However, don't position shelves above the lavatory or bidet if people could knock their heads. Sometimes it is possible to fix one long shelf level with the top of the cistern, but don't enclose this – it must be easily reached for any plumbing emergencies.

If the linen cupboard is in the bathroom along with the hot-water tank, this is an ideal place to air linen on slatted shelves, and to store dry goods which won't perish in consistent warmth. However, linen should not be stored permanently in a warm atmosphere, so try to provide storage in a separate room for towels, sheets and pillow cases. Remember to allow for the different sizes of folded sheets and duvet covers.

If you build in the basin, perhaps with a vanity unit, or simply conceal some shelves below the basin behind a curtain, this will make a much neater-looking feature as well as providing more storage space (this can also be done in bedrooms where there is a basin). A 'skirt' covering a table can hide stacking crates or basic shelves.

Above: *The simplest linen storage – the traditional cane basket.*

Left: *A pull-out bin for linen can be part of a run of bedroom cupboards, or built into a bathroom unit.*

BEDROOMS

Most bedrooms are used to store clothes, personal effects and luggage, and when the room has to be dual-purpose you may have to find a place to store hobby equipment, books, papers and, in the case of children's rooms, toys and sports equipment.

The most practical ways of providing adequate storage include running fitted wardrobes or cupboards along one wall; filling the recesses each side of the fireplace; and, in a smaller room, creating an alcove for the bedhead by placing a wardrobe to each side of the bed, linked with top cupboards above. These arrangements can be streamlined still further by running units along under the window to provide a desk or dressing table as well as storage, building in a window seat with lift-up top, or placing freestanding furniture here.

Whether you plan to get custom-made built-in furniture, freestanding furniture or a do-it-yourself installation, remember to measure the width of bulky items on hangers and the lengths of overcoats, long dresses, jackets, shirts and blouses. Rails, shelves, and pull-out drawers can then be tailored to fit. An inexpensive way of storing clothes is by hanging them on a dress rail – this can be the collapsible chrome type, or you can fit a wooden or metal one in a recess – which can be hidden behind a curtain, screen or pull-down blind (shade) which co-ordinates with the rest of the colour scheme. Don't forget to make use of the space underneath any hanging area for shoes – proper shoe racks are often the best way of dealing with this. Any short hanging space (for jackets or blouses) can have pull-out wire baskets.

Below: *Do-it-yourself storage system from a builder's merchant – you can buy the pre-cut pieces and build it yourself. Plan your storage requirements first, and work from the inside out when custom-building wardrobes.*

Above: *Fitted cupboards above and to either side of the bed create a cosy alcove around the bedhead and make the most of available storage space.*

CHILDREN'S ROOMS

Children's rooms usually have to be the most flexible of all the areas in the home – and consequently need very careful planning. Generally the room starts off as a nursery for a baby – used only for changing, bathing and feeding the baby, and storing their clothes, nappies and toiletries. But all too soon it becomes a dual-purpose room, doubling as playroom and bedroom; then when schooldays start it may have to combine sleeping, play and study facilities. At some point it may have to be shared with another child, and you may wish to provide some private areas for each of the occupants. The final stage could be a teenager's private room, or even a one-room living area for a young couple.

A ROOM TO GROW WITH THE CHILD

Design the room so it can adapt to the changing requirements of the child and start off with a few basic items which can be added to as the child grows and needs more storage space. The main surfaces should be tough and easy to clean to withstand rough treatment. Colour schemes can be bright, cheerful and stimulating. Any currently fashionable favourite nursery, cartoon and television characters are best confined to wallcoverings, borders, bedding and posters, which are relatively easy and cheap to

change as one fad goes out and the next one comes in.

So start off with the basic nursery. Although some of the items will be child-size, (crib, cot, small table and chairs, low bookshelves, first desk) don't use scaled-down storage cupboards and units – just adapt the interiors of regular-sized furniture to take small garments, and alter the hanging space, internal drawers etc as storage requirements change. Try to choose from a 'stock' range to which you can add extra pieces later, or buy and paint some old furniture, which will allow for extra additions which can be painted to match.

TOY STORAGE

Expand the facilities at toddler stage, with the addition of plenty of play space and toy storage. It is often possible to use the area under the window for a toy box with a lift-up top, which can be padded to make a window seat. Or position two low-level storage units under the window, leaving a 'kneehole', and link these with a play top, which can later be used for a desk.

Other practical ideas include wheeled toy boxes, or coloured stacking crates, to fit into cupboards, under desks or under 'skirted' tables. Wheeled trolleys can also be a useful device – old wooden or tubular metal ones can be brightly painted to match the room; or for a more streamlined look, get ideas from

hospital and catering catalogues. Don't put such furniture into very young children's rooms, as you may well find they use it as a vehicle to ride about on.

It is sometimes possible to incorporate storage into the bed. Some come ready-made and are called cabin beds – ideal if you want to give the room a nautical flavour. In a simpler theme, pull-out drawers, or some of the items suggested above, can fit underneath an ordinary bed, where often there is wasted space. It is sometimes possible to build a higher-level bed, and combine it with storage underneath, which might be decorated to look like a doll's house, a toy theatre or cinema, a fort, a railway station or a garage.

SPECIAL REQUIREMENTS

Split-level bed treatments are even more practical when a child reaches school age and has to do homework – a bed on a platform, reached via a ladder, can have a wardrobe and desk underneath, and a set of wall-mounted shelves. In a large room it could even include a shower cabinet or basin/dressing table, while the desk might go under a window. This could be a boon in a family where there is a great demand in the morning rush for bathroom space.

Some bunk beds can be positioned at right angles to each other, with space below one and above the other for storage. Such an

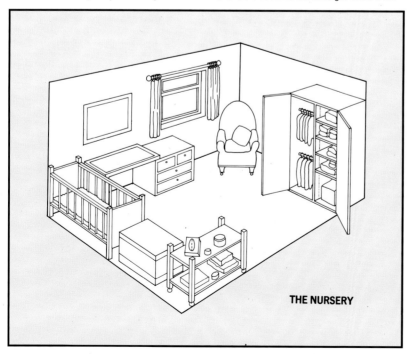

THE NURSERY

Left: Stage 1: the nursery. A large cupboard (freestanding or built-in) can have adjustable shelves or baskets in one side for small garments, linens and some toys, with hanging space on the other side. A two-tier rail (one above the other) will allow for hanging small garments; a changing area can be created by using storage units, or a chest of drawers and cabinet linked by a worktop with changing mat. A cot, trolley for baby toiletries and nursing chair complete the scene.

Right: *Stage 2: toddler's room. Remove the nursing chair; remove one dress rail and add extra shelves in cupboard; retain the cot and toy box; add a blackboard or other drawing surface; put a pinboard above the play-and-painting surface (which was the changing table); add a small book case or bookshelves.*

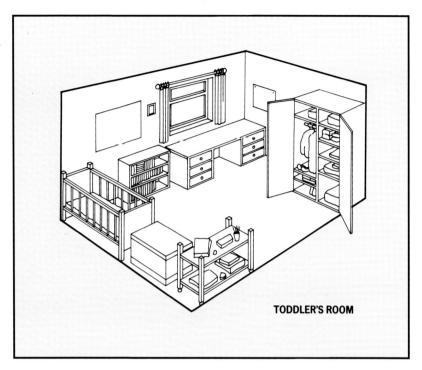

TODDLER'S ROOM

Above: *Sophisticated custom-built storage in a child's room provides a touch of fantasy. The bed is built over a doll's house, toy theatre, book shelves and toy storage.*

DECORATIVE STORAGE

You can make decorative use of a necessity in the bedroom. Accessories can look very attractive displayed dramatically – long beads draping a mirror; small pieces of jewellery, or colourful cotton wool puffs stored in a big glass jar; fans in a large pottery jug; rings on a plaster cast of a hand; brooches pinned to a piece of framed felt hanging on the wall; belts suspended from a decorative rack; scarves swathing a screen. Cosmetics can look pretty in baskets – some of those intended for cutlery can look very decorative, especially if you paint the wicker in a pastel colour. In a child's room, under-bed storage can be decorative, practical and fun (*see* right); incorporating room for books and toys and a place for a toy theatre.

arrangement can be made from brightly coloured scaffold-type tubing, specifically sold for the purpose, and in some cases in kits. And as the child reaches the teen years and may want a 'hi-tech' bedroom, industrial shelving can be adapted to many practical purposes.

Clothes, sports and hobby storage are covered elsewhere in this book, and can be adapted to children's needs, but there may be a need to accommodate a special hobby. Firm but collapsible trestles can be combined with a stout top for a train or car layout; a large cutting-out, modelling or painting top – even a ping-pong table. The trestles can be hung on stout hooks on the wall when not in use, and if the top cannot be stored elsewhere in the house, it might be raised to the ceiling on a pulley system, or flap back against the wall on hinges. Obviously such arrangements are not suitable for very young children, and must be tested for complete safety – and checked regularly afterwards.

DIVIDE AND RULE

When a room has to be shared by two or more children, it is essential to try to provide some personal space and privacy. It is sometimes more practical to give over a larger room (the main bedroom or a downstairs dining room) to them, so there is enough room to divide the area adequately. You can often do this by placing the storage units at right angles to the walls, to create natural divisions without resorting to any structural alterations. These can also be rearranged as the child grows.

One problem is often lack of light in part of the room – try to spread the natural daylight as much as possible. This can sometimes be done with mirrors, or replacing panels in doors with panes of glass, or making an internal window. It is also essential to plan adequate artificial light on separate circuits so each area of the room can be controlled individually. If you divide the room by placing a wardrobe or cupboard at right angles to one long wall, with a lower chest, desk or unit beside it, a second tall storage unit can complete the run of furniture, forming the dividing line, and positioned to face into the

opposite part of the room, so the storage can be shared equally.

The backs of such cupboards are usually not pretty, but they can be decorated to match the room, with fabric, felt, paper or whatever. This then becomes the ideal place for one of the bedheads, with a wall-mounted shelf, or the top of the lower dividing unit, used as a bedside table. The space above could be closed off with a curtain, roller or

Left: Decorative pegs are ideal for hanging small items in children's rooms, and are inexpensive to replace when the child grows.

vertical blind hung from the ceiling.

If this is not feasible, then open shelving, which can be reached from both sides, above more solid lower units or chests can provide storage and work as a natural divider. If you do this, take essential safety precautions and make sure the furniture is firmly fixed and the upper layer cannot topple over.

In smaller rooms where such splitting is not feasible, the furniture may have to be ranged round the walls; or consider lateral divisions, with beds on platforms as previously described. Again, a floor-to-ceiling blind or curtain can help to create some privacy and be opened when the entire space must be used as one room. Flexible, temporary divisions are best in children's rooms, when requirements and interests are likely to change every few years.

Right: Stage 3: school child's room. Remove the cot; alter hanging space for larger clothes by removing shelves in the wardrobe; put toy storage in the bottom; convert the play top to a desk; add bunk beds.

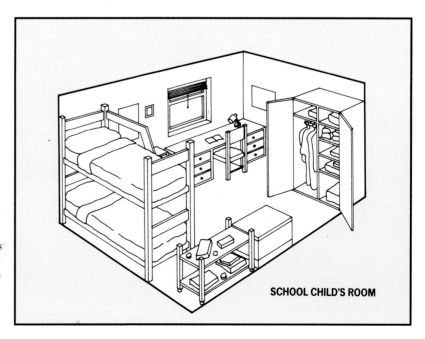

SCHOOL CHILD'S ROOM

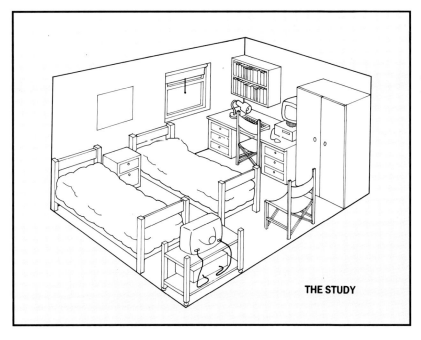

THE STUDY

Left: *Stage 4: the study. Split the bunks, place a bedside table between them; add more shelves above the desk; add computer; use the trolley for a portable TV; add a pair of folding film-director style chairs for visiting friends (which could be hung on the wall when not in use).*

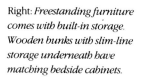

Right: *Freestanding furniture comes with built-in storage. Wooden bunks with slim-line storage underneath have matching bedside cabinets.*

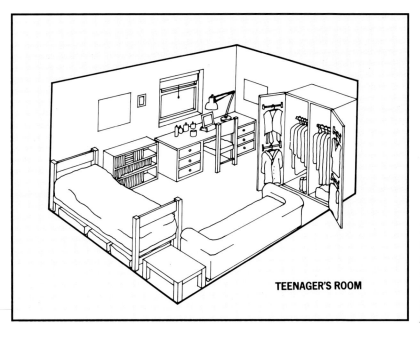

TEENAGER'S ROOM

Left: *Stage 5: teenager's room. Remove bunks and replace with a single bed and a sofa bed; place storage units under single bed; convert wardrobe to all-hanging space; replace trolley with coffee table; make dual-purpose desk/dressing table with freestanding or wall-mounted mirror.*

FOCUS ON SHELVING

Shelving comes in many different forms, from purpose-built timber or 'instant' laminated shelving to delicate light-filtering glass. Shelves need not be merely utilitarian, used just to provide storage: they can provide the perfect place to display an attractive collection of objects, glass, books, or china. The shelving system may become an integral part of the room scheme – and should be in keeping with the architectural style, or the ambience which you are trying to create.

There are so many units and systems available that you are certain to find one to suit your budget, from adjustable systems to 'instant shelving'. Bear in mind the size and weight of the objects you want to display will to some extent dictate the size, depth and strength of the shelf and may also influence the type of shelf and the type and method of fixing.

Above: *Shelving need not be expensive, yet can still look stylish in any room setting. Lamps are angled to accent objects.*

SUITING YOUR SCHEME

In a traditional-style living-room, for example, purpose-built painted shelves for books and ornaments could be slotted into a recess above matching panelled cupboards, trimmed with beading. But in a hi-tech setting, similar storage might be provided by heavy-duty industrial shelving that clips onto batten strips, with clip-on wire baskets, racks for records, and cubed pigeon-holed boxes set amongst them. Such streamlined systems suit this no-nonsense decorating style.

If you have a window which does not have to be opened or draped, you can convert it into a display showcase by 'stretching' glass shelves across the reveal. Use special 'invisible' brackets which are not noticeable through the glass, or use decorative ones (like curly wrought iron) which are designed to be seen and become part of the overall design. Stand plants or items of coloured glass on these shelves, so when the light filters through, you get dramatic shadow-play, and interesting coloured shapes projected onto the floor or walls. Such shelves should be lit at night for maximum effect, either from above, below or to the side – or with integral lighting fixtures.

You can also make an attractive display out of a necessity in kitchens and bathrooms. In these rooms you often need to store relatively small items, which are best put on shelves, and which can look very decorative if they are out on view. In a modern kitchen, for example, you might use metal or glass shelves to hold streamlined kitchen equipment, or pulses, colourful pastas and herbs in glass jars, but in a country-style room, pine shelves, dressers or plate-rack shelving would be more in keeping, used to display china, pottery, pewter, flowers and glass as well as the more mundane items.

In the bathroom, soaps, bottles and jars containing various lotions can all add colour and personality to the scheme and, when combined with shells, steam-loving plants and other interestingly textured items, can become a decorative focal point if displayed on suitable shelving. Again glass or acrylic shelves can look good in this situation – or tiled ones to match the bathroom scheme are practical to keep clean, but they will be heavy so make sure the brackets are strong enough. Shelves are not always considered necessary in the bedroom, unless it is a child's or teenager's room, or a dual-purpose area, where they are obviously essential to store

---

BUYING GLASS SHELVING

If you are ordering glass shelves, most glass suppliers will cut them to order for you, and will advise you on the correct thickness of glass to get according to the weight of the objects you wish to display.

Always have all the edges of any glass shelf polished so there can never be an accident when handling or cleaning them.

If you are worried about possible breakage you could use clear acrylic (Perspex or Plexiglas) shelves, but they are more expensive than glass and do tend to scratch. They are also more flimsy, but you can get some fabulous colours in clear plastic.

Right: A simple wooden shelf above a tiled border in a country-style kitchen is used for storage and display.

books, files, toys, items required for hobbies, records, cassettes, and video tapes. But there is no reason why the bedroom should not have attractive display shelves – or practical ones for books and files.

Bedroom recesses can be filled in a similar way to those in the living-room, and sometimes, in older houses, shelves can bridge the gap between fitted cupboards built into alcoves to each side of a fireplace. If the recesses are not deep enough, and the cupboards have to project into the room, shelves mounted across the breast, linking the wardrobes can help to achieve a balance and neaten the overall effect. If the fireplace is no longer operative, and the grate and fireback have been removed, you could use the opening for fixed or adjustable shelves.

To be practical, the chimney must be either capped or have a cowl put on the top (so rain, snow or birds can't drop down inside the opening), and some form of 'throat' be fixed across the top of the fireplace opening. A source of ventilation may also be necessary, so the wall does not become damp. The resulting alcove is often a good place to put a music system or a television set, and can be fitted with doors to hide it all away. But if you live in a built-up residential area with many houses, think of the neighbours as noise will travel through the disused chimney into the next-door property.

If there is a basin or vanity unit in the bedroom, or the dressing table is fitted into a recess, shelves can be positioned round a mirror in a similar way to bathroom shelving.

PLATE DISPLAY SHELVES

1 Hold the moulding temporarily on the shelf to find the optimum distance which will prevent the plates from tipping but will keep them as upright as possible for display purposes. Mark the position of the back edge of the moulding on the shelf. Tap in panel pins through the face of the moulding to just protrude on the underside – approximately 12 mm (½ in) should ultimately enter the shelf – one 32 mm (1¼ in) from each end and the others at approximately 20 cm (8 in) intervals.

2 Apply adhesive (white glue) to the underside of the moulding and pin in position against the marked line. Wipe off excess adhesive with a damp cloth.

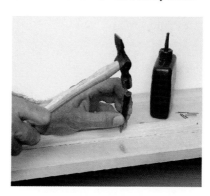

▲ 3 Punch the pin heads below the surface with a nail punch and fill with cellulose filler (if painting) or matching wood filler.

Above: *Timber can be used to run around the perimeter of a kitchen, dining-room or hall wall at picture-rail height. It may be fixed to shelving to create a plate restraint for display purposes.*

MOULDING FOR PLATE DISPLAY SHELVES

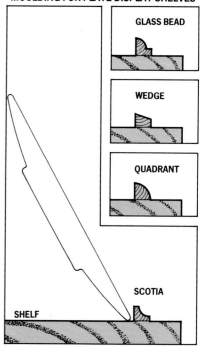

GLASS BEAD

WEDGE

QUADRANT

SCOTIA

SHELF

Above: *This diagram illustrates the various timber moulding sections that would be suitable for use as plate restraints.*

SPICE RACK

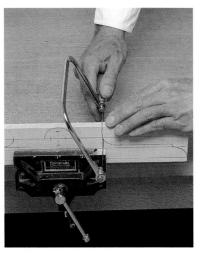

▲ 1 Cut the end pieces to equal length (about 30cm/12in, depending on jar height and allowing clearance above the tilted jar). Cut the shelves and mark out the shelf positions according to the jar height(s). The lipping on the top shelf should finish flush with the top edge of the side pieces.

▲ 2 Mark out and cut the scrolled top piece, using a coping saw or a jigsaw with a scrolling blade (the ends should finish flush with the top edge of the side pieces). Sand smooth.

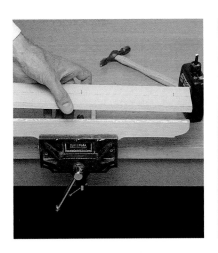

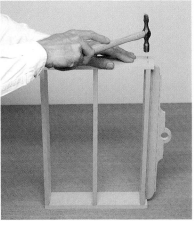

▲ 3 Mark the centre line along the underside of the top shelf. Hammer in 25mm (1in) panel pins along this line, to secure the top piece, until the points just protrude. Hold the top piece upside-down in a vice, and apply PVA woodworking adhesive (white glue) to the bottom edge. Pin the top shelf to it, centring the pins in the top piece.

▲ 4 Tap 38mm (1½in) panel pins just through the side pieces, across the centre line of the shelves. Apply PVA woodworking adhesive to the ends of the shelves, and pin them in position; wipe off excess adhesive with a damp cloth. Punch the pin heads below the surface and check that the unit is square.

▲ 5 Cut the plywood back panel to size – full width and from the underside of the bottom shelf to the top of the top shelf. Glue and pin it in position with 12mm (½in) panel pins. Cut the moulding to length – the full width of the unit. Glue and pin to the front edge of each shelf with 19mm (¾in) panel pins, flush with the underside and overlapping the side pieces. ▶

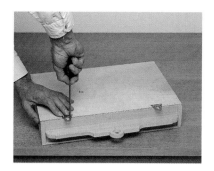

▲ **6** Fit slotted mirror plates to the back of the unit, screwing through into the top shelf. Measure the hole spacing and drill and plug the wall accordingly. Screw in 32mm (1¼in) roundhead screws and test-fit the unit. Check that it is level and secure. Adjust the screw tightness as necessary. Remove the unit from the wall and paint or varnish it.

PLATE RACK

▲ **1** Mark out the shelves together, to ensure that they are exactly the same length, and cut to length. Label them 'top', 'second' etc, and label the ends 'left' and 'right'. Cut the sides to length and mark the shelf positions (according to the sizes of plates) and the positions of the plate rests. Again, mark both sides together to ensure that the shelf and rest heights are identical. Remember that one side piece will be a mirror image of the other. Cut the rest dowels to the length of the sides plus 25mm (1in).

▲ **2** Mark the centres of the shelf fixing dowels on each end of each shelf, centred 19mm (¾in) from the front and back edges. Drill 6mm (¼in) holes 25mm (1in) deep using a dowel bit and dowelling jig. Allowing for the thickness of the mirror plates at the back edge of the top and bottom shelves, mark corresponding dowel holes on the sides and drill 15mm (⅝in) deep.

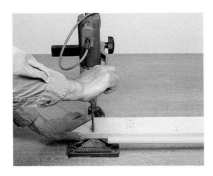

▲ **3** Drill 22mm (⅞in) diameter holes 12mm (½in) deep in the sides for the plate-rest dowels, so that the front of each dowel will be 4mm (³⁄₁₆in) from the front edge of the sides. Use a flat bit or end mill in an electric drill with a depth stop, and be careful not to drill right through.

▲ **4** Screw mirror plates to the back edge of the top and bottom shelves, 75mm (3in) in from each end and in the centre, with single hole uppermost.

▲ **5** Glue and pin half-round moulding to the front edge of each shelf, flush with the top surface.

◄ **7** Glue and pin D-moulding to the front edge of each side and hang the unit on the wall by screwing through the mirror plates into the appropriate wall fixings.

Below: *Plate racks along shelves will enable you to make a decorative feature out of kitchenware and are particularly suited to country-style kitchens.*

▲ **6** Glue dowels into the shelf ends and tap home with a mallet. Wipe off excess adhesive. Assemble the shelves and rest dowels to one side piece, with PVA woodworking adhesive in the dowel holes and across the shelf ends. Tap home with a mallet. Fit the other side piece in the same way and cramp the assembly together with sash cramps (if you have or hire them) or with Spanish windlasses. Make these by tying stout cord round the unit (with padding between the cord and the wood) and twisting the cord tight with an offcut of timber. Check that the unit is square.

ALCOVE SHELVING

Above: *It is simple to fix wooden shelves into an alcove to make the most of otherwise wasted space. Here, the shelves were fitted either side of a fireplace and painted to blend into the background.*

△ **1** Cut 50 × 25mm (2 × 1in) back batten to length, 10mm (½in) or so less than the width of the back of the alcove at the required height. The gaps will be concealed by the side battens and prevent problems with rounded or unsquare corners. Hold batten at required height on wall, with spirit-level on top, set level, and mark along top edge onto wall.

△ **2** Mark fixing hole positions on the batten 40mm (1½in) in from each end and at 40cm (16in) intervals between them. Drill clearance holes for screws and countersink the holes so that the screwheads will lie just below the surface. A minimum screw length of 45mm (1¾in) will be needed. Hold

batten, top edge against the marked line, and mark through the fixing holes with a nail. Drill the wall for wall fixings, so that at least 25mm (1in) of the screws is in brickwork, insert the fixings and screw the batten firmly in position, checking that it is level.

▲ **3** Cut the side battens to length, allowing for the thickness of the back batten, so as to finish 6mm (¼in) from the front edge of the shelf. Angle the front ends to make them less visible. Drill fixing holes 40mm (1½in) in from each end and countersink them. Hold each side batten in position, butting up against the back batten and with a spirit-level along the top edge, and mark through the fixing holes. Drill and plug the wall and screw the side battens in position.

▲ **4** Measure the depth of the alcove and deduct from this the thickness of the front reinforcing batten (see step 6). Cut the shelf to this width. Two thin battens, each shorter than the width of the alcove and with one end cut at an angle, can be used to measure the internal dimensions of the alcove. Hold the battens together and overlapping, pointed ends outwards, along the back batten and slide them apart until the ends just touch the side walls of the alcove. Clamp together.

▲ **5** The corners of an alcove are rarely square. Before transferring the width of the alcove to the shelf, use a try square and steel tape measure to work out whether they slope inwards or outwards from the back corners. Next transfer the distance between the ends of the clamped battens directly to the shelf, leaving about 25mm (1in) of waste at one end for cutting any angles and allowing 2–3mm (⅛in) clearance. Using your try square from the back edge of the shelf, mark on the calculated deviations from square and double check the marking by using the pair of battens across the front of the alcove. A final complication: if using a jigsaw to cut the shelf to length, mark it on the *underside* and cut with this side upwards. The saw cuts on the upstroke, so damage to the surface will be hidden when the shelf is in position on the battens.

◀ **6** The shelf should be a close fit to the shape of the alcove. Any trimming necessary can be done with a block plane. The shelf can be secured to the battens with small angle brackets. To prevent the front of the shelf from sagging under the load, glue and pin a batten along the front edge. Punch the heads of the pins below the surface and fill with matching wood filler or cellulose filler (if painting).

Above: *Glass shelving in a recess is illuminated by a recessed downlighter.*

INTEGRAL LIGHTING

Various ways of illuminating shelving will be discussed and illustrated in the lighting section, and the method you choose will depend on the type of shelving you install; the objects you place on the shelves – and whether you want to accent them, or merely to be able to see them clearly. Don't forget to allow for this in your basic planning, so any necessary wiring can be done at the construction stage.

One of the best ways of lighting display shelves, or those which you want to create a warm background 'glow' in the room, is by integral lighting which is fixed to the shelf, or built in as part of the shelving system. Low-voltage tungsten lights housed in slender filament tubes, or the transparent flexible tube with encapsulated bulbs can be fixed to the back or front edge of a shelf, and concealed behind a neat facing strip if necessary, to avoid glare. Some ready-made display units come with a system built in.

Downlighters or spotlights can be fitted into the top of fixed shelving, or set into the ceiling above any wall-mounted shelving in recesses; a strip light can be hidden behind a 'pelmet' or cornice going across the top; and ceiling-mounted wall-washers can bathe a set of shelves with soft light. Uplighters can be positioned in the bottom of some fitted shelving, and can look particularly effective when the light is shining upwards through glass shelves.

Alternatively you can place portable lamps on shelves (including uplighters), and some of the shadeless ones will illuminate all the items on the shelf clearly. Spotlights clipped to the shelves or on track can also be very effective. Attach the track to the ceiling above a bank of shelves or, for greater flexibility, position it vertically down one, or both, sides of the shelves. The spots can then be angled strategically to light various points of interest, and altered when you want to change the arrangement.

CHOOSING SHELVING TYPES AND MATERIALS

Having worked out where you will have the space to site shelves, and what you want to place on them, you will need to decide what they are going to be made from, and how they are going to be supported. This in turn will relate to the size and weight of the items you want to store. Also decide whether they will be fitted and fixed, or flexible and adjustable – you might buy the former ready-made as an item of furniture, or have the system tailor-made to your requirements, although this type of shelving can be a do-it-yourself project.

Left: *Adjustable shelving in a rich mahogany-look finish creates a warm, welcoming look in an Edwardian-style living-room.*

ADJUSTABLE SHELVING

Left: *Adjustable shelving is fitted into a living-room chimney breast recess, and can be adjusted to accommodate different sized items as necessary. It is coloured to contrast with the background and form a design feature.*

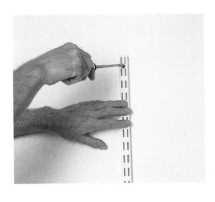

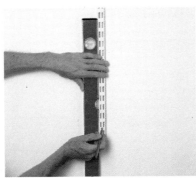

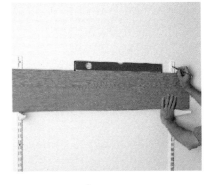

▲ **1** Determine position of shelves on wall and mark first upright position about 100mm (4in) from end of shelves. Mark through top hole of upright. Drill hole for fixing according to screw size. Hole should penetrate through plaster – approximately 12mm (½in) and at least 25mm (1in) into brickwork. Insert wall fixing and screw through top of upright. Don't tighten completely.

▲ **2** Set upright vertical using spirit-level, and mark through bottom hole onto wall. Swing upright to one side, drill and plug as before, and screw in position. Again, don't fully tighten. Mark through remaining holes, remove bottom fixing screw and swing upright aside. Drill and plug remaining holes, then screw firmly into position through all holes.

▲ **3** Insert a shelf bracket at a convenient height in the fixed upright and a second bracket in the corresponding slot of another upright. Mark the approximate position for the second upright, the same distance in from the end of the shelves. Rest a shelf, on edge, across the two brackets, hold a spirit-level on the top edge, and raise or lower the second upright until the shelf is perfectly horizontal. Mark through the top fixing hole of the second upright. Screw into position as with the first upright.

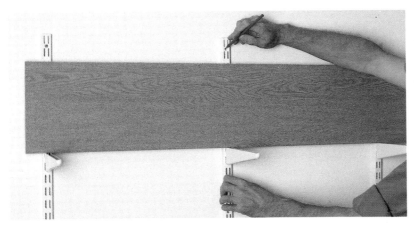

◀ **4** Position intermediate uprights by again inserting a shelf bracket in the corresponding slot to those in the outer uprights and bringing it into contact with the underside of the shelf on edge. Screw them into position as before. The spacing between uprights should not exceed the maximum support spacing for the shelf material (*see* table on page 45). Finally insert brackets at the required heights in all the uprights and rest the shelves on them. If there will be young children around, screw through the brackets into the shelves.

Above: This teenager's room is geared to the changing requirements of its owner. The flexible storage system is used to support wide shelves for desk top, computer and VDU and to divide the sleeping area from the working part of the room. A filing cabinet completes the storage facilities.

READY-MADE ADJUSTABLE SYSTEMS

The flexible method is very simple to install, since it consists of metal uprights of various lengths from around 37cm (14in) to 2.4m (8ft), which can be cut to size with a hacksaw. They can be anodized silver and gold, or enamelled black, brass or bronze. Some systems are made in white or a coloured satin finish and you can select a colour so the upright blends in with the background, or so that it stands out if you want to make a feature of it.

The uprights are fixed vertically to the wall and, in a normal recess, two or at most three are sufficient. But with a wider span – or if you are supporting very large items and heavy shelves – you will need more of the upright struts and brackets. Manufacturers make recommendations which are available at the point of sale, so check this when buying the system. The brackets fit into the slots in the upright struts, which are screwed to the wall. The shelves are supported on these brackets, which can be moved up and down the

uprights making the shelving flexible and able to be adjusted to your requirements.

The slots (and corresponding fixing end of the bracket) can be square, oval, T-shaped or double. It is essential to buy the correct brackets for the uprights, as you can rarely marry together two systems. One type has a continuous channel which allows the brackets to be fixed at any point.

The shelves themselves can be made of a variety of materials from solid wood and melamine-faced chipboard to glass, and the brackets will need to be the right type to support them – again weight will be a consideration, as well as the depth and length of the shelf. The brackets need to be almost as long as the depth of the shelf (the front edge of the shelf should not project more than 25mm/1in beyond the end of it). In some cases the shelf is actually held by a groove in the front of the bracket; in others the shelves are actually fixed to the bracket.

There are a limited number of adjustable timber systems available and there is an

alternative metal system which uses clips instead of brackets. The strips are available in bronze- or zinc-plated steel, which is cut to size and four lengths are screwed to the sides of an alcove, or to the side panels of a bookcase (these can be rebated into the side panel). Bookcase clips slot into the strip and come in a variety of shapes and sizes. Some are designed for heavy shelves which makes them practical for use in a study or home office, and other clips are suitable for use with glass shelves.

PANEL-MOUNTED ADJUSTABLE SHELF SUPPORTS

It is possible to fix shelves in place by means of pegs or dowels (which come in various shapes and sizes in metal, plastic or wood) knocked into similarly sized holes spaced at regular intervals along the length of a wooden upright. Four pegs are inserted at each level to sit under each corner of the shelf for an invisible fixing, and special pegs or clips are also available for glass. You can cut wooden dowelling to length and tap it into the holes to support the four corners of each shelf.

This system also operates using a wire support, which has the advantage of being almost invisible. Holes are drilled in the uprights to hold the ends of a shaped length of 3mm (⅛in) galvanized wire; the ends of the shelves are grooved and slide over the wire which supports them horizontally at each side. These methods are not suitable for heavy weights, but there is a two-part metal clip which can be used, which works on a principle similar to a hinge and takes heavier shelves and contents.

INSTANT SHELVING

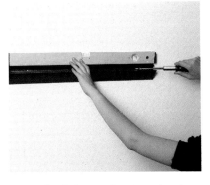

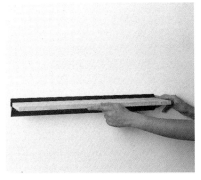

▲ **1** Check the true horizontal, using a spirit-level, and mark the wall. Check again for accuracy when the support is in position, and screw to the wall.

▲ **2** Position the shelf in the slot of the support, with the front edge higher than the back edge, and push inwards and upwards to engage it. Test for rigidity and strength before positioning items selected.

Right and above right: Simple but strong supports can be used to fix shelves to the wall. They can take any type of shelf, and are particularly good for glass as the fixings are almost invisible. They are also good for streamlined storage in a modern interior, or to wall-mount equipment and tidy away tools and paint tins in a garage or workshop.

WALL-MOUNTED FIXED SHELVES

With the fixed method, the shelves are attached permanently to the wall by means of strong shelving brackets or supports, which can be made of metal, plastic or wood – some are suitable for a flush wall and others fit into an alcove. It is advisable to secure the shelves by attaching them to the brackets, and again you should choose your bracket in relation to the size of shelf and the weight which it will have to bear: more brackets will have to be fitted across the width of the wall for heavy loads (*see* chart opposite).

Some brackets are L-shaped – made of metal, steel or aluminium in sizes from 15cm (6in) to 60cm (2ft) to fit standard shelf widths. Again the front edge of the shelf should not project beyond 25mm (1in) of the bracket tip, so choose a suitable size and thickness. Some of these brackets are made in a decorative curly design, and can look very attractive when combined with transparent display shelves.

Triangular brackets are only used for fitting a shelf into an alcove, and are usually more utilitarian in design, and made of steel. One part of the triangle is fixed to the wall and one supports the shelf, helping to distribute the weight evenly.

Cantilever brackets are metal rods which are inserted into a hole in the wall and into the back edge of the shelf at each end. They come in a variety of different sizes and are made in metal or impact-resistant plastic and are moulded to a high-tensile steel pin to provide extra strength.

Wooden battens can also be used to support fixed shelving. Lengths of wood are fixed to the wall to support the edges of the shelf and are only suitable in a recess or alcove since they have no supporting arm. The thickness of the battens will depend on the weight of the shelf and the load. They are usually available in 25 × 25mm (1 × 1in), 38 × 25mm (1½ × 1in), and 50 × 25mm (2 × 1in) sizes, but alternatively you could use pieces of wood cut to a suitable size for the intended purpose. (*See* pages 38–39 for details on installing shelving in a recess.)

There are also angled metal strips and high-strength continuous aluminium brackets, which are basically L-shaped. The shelf sits within the 'L' so must not fit too tightly against the wall. The continuous strip type support the shelf along its full length and take 15mm (⅝in) or 18mm (⅘in) board or 6mm (¼in) glass shelving. Lengths range from 60cm (2ft) to 2.4m (8ft).

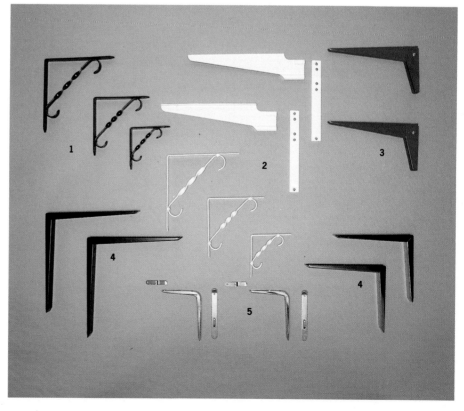

Right: *Shelves can be fixed to the wall using a variety of brackets – select them to suit the situation, style of the room, and weight they have to take. Examples illustrated include decorative steel brackets with wrought-iron styling in a range of sizes, epoxy-coated in black or white (1); heavy-duty steel construction cantilevered, mitred and concealed fixing brackets (2); folding shelf brackets (3); medium weight pressed steel right-angle brackets (4); and brass-finish brackets for a more traditional look (5).*

Above: *A space-saving bracket with a unique folding action in use. It is designed to support shelving up to 45cm (18in) deep.*

FIXINGS

Most battens, brackets or uprights intended for holding shelves will have to be fastened to the wall by means of woodscrews, and you will need to select the right length and gauge according to the shelf, its support and the load it has to bear. The type of wall will also determine your choice; for example, woodscrews will not grip into plasterboard or masonry satisfactorily so you will need to fit a wall-fixing device into these materials first, before inserting the screws.

In solid walls you can use a general-purpose plastic plug, a heavy-duty plug, a light-weight cellular block plug or – for securing heavy loads to masonry or concrete – you will need a masonry bolt. For hollow walls there are various special plugs and toggle bolts. These have barbs, wings or toggles which spread out behind the board once they are in position, evening out the load behind the board.

For most shelves 50mm (2in), 62mm (2½in) or 75mm (3in) woodscrews will be suitable. In solid walls the screw must be long enough to pass through the support and at least 25mm (1in) plus the plaster depth into the masonry. In hollow partition walls the length of the screw will depend on the type of wall fastening being used – some of those described above come complete with screws.

SHELVING MATERIALS

There are some plastic and laminated ready-made shelves available. These, and some other forms of shelving material, come ready-finished in standard sizes, and some come ready-coloured. If you want colourful shelves there is no reason why you should not paint or stain the material. Almost anything can be used for the shelves themselves, with a wide variation of look, quality and price. These can be purchased in lengths, but you can have them cut to size, so make sure your measurements are accurate.

Choose from melamine-coated chipboard (particle board) heavyweight or wood-veneered chipboard, softwood, hardwood, medium-density fibreboard or plywood (which will need varnishing), Perspex (Plexiglas) which must be at least 10mm (⅜in) thick, or plate glass (which should be at least 6mm/¼in thick, but selected according to the weight it will have to bear). Remember, some types of shelving (like glass) will need special brackets, so it makes sense to decide on the type of shelving first.

LOAD-BEARING SHELVING MATERIALS

The following chart indicates the maximum span allowed between supports for medium to heavy loads. Bear in mind that if glass shelving is used, the shelves themselves will be relatively heavy. Use the correct brackets, and ensure as with all shelving that supports are firmly secured.

material	thickness	length between supports
blockboard	12mm (½in)	45cm (18in)
(lumbercore board)	19mm (¾in)	80cm (32in)
	25mm (1in)	100cm (39in)
faced chipboard	15mm (⅝in)	50cm (20in)
(particle board)	19mm (¾in)	60cm (24in)
	25mm (1in)	75cm (30in)
plywood	12mm (½in)	45cm (18in)
	19mm (¾in)	80cm (32in)
	25mm (1in)	100cm (39in)
timber	15mm (⅝in)	50cm (20in)
	22mm (⅞in)	90cm (36in)
	28mm (1⅛in)	106cm (42in)
glass	4mm (⁵⁄₃₂in)	Note: Glass is not suitable for
	6mm (¼in)	heavy items, and the shelf
	10mm (⅜in)	span must be calculated in
	22mm (⅞in)	relation to the load.
		A maximum span of 50–60cm (20–24in) may be advised, but check with a supplier.

DECORATING SHELVES

Fixed and adjustable shelving systems can be made from many different materials (*see* pages 40–45), but in the main, wood or some type of board is chosen for the shelves themselves, to slot into one of the various fixing systems. This gives greater flexibility for decorative treatments. The shelves can be coloured to 'fade into the background' – paint or stain them and their supports the same colour as the wall or recess behind them. Or they can be styled to stand out, so they become a decorative feature in their own right.

Stain them to match dark wood furniture in a traditional living room, for example, and position against a pale apricot, pink or sky-blue wall area; treat them to look like antique pine and show them off against whitewashed walls in a kitchen setting; stain or paint shelves a rich foliage green and support them on white ladders, lattice or trelliswork in a conservatory or garden room; paint the shelves yellow and the struts primary blue and set them into a paler blue recess – or work up a similar theme with jade green, terracotta and aquamarine to brighten a dark dining area.

In a modern living room with a metropolitan theme, install black ash shelves to complement matching units, suspended on silver anodized aluminium struts, and paint the surrounding area in a bold primary colour, or use a deep-coloured hessian wallcovering. Use Scandinavian-style light teak shelves on brass struts, or on black or white ladders to fill a triangular space under stairs. If white melamine or laminated chipboard (particle board) shelves are selected for a modern setting, a teenage bedsitting room or child's room, choose colourful struts, slotted angle or metal tubing supports to co-ordinate with the rest of the decorating scheme.

Above right: Add an extra decorative touch to a freestanding pine dresser – edge the shelves with a paper or fabric border to link with the theme of the china or room scheme.

Right: Decorative edging on small neat shelves can be simply shaped with a fretsaw.

EDGES

Don't forget to take any colour or stain onto the edge of the shelf. In some cases this can be difficult – the stain or paint is 'swallowed up' by the core of the board, and the result looks grubby. A finishing strip may have to be added to the front edge to take up the colour to the same degree as the rest of the shelf. But if this is not feasible, there are many attractive ways of finishing off the edges of shelves to create a decorative and highly individual look.

The treatment you select will depend on the room and its styling; for a feminine or country cottage effect, a neatly frilled fabric valance (made in the same way as a curtain) can be pinned to the edge of the shelf, or suspended on curtain wire, slotted through a cased heading in the fabric, and screwed to the edge of the shelf in several places. This looks crisp and fresh in a kitchen setting if the fabric is red and white or blue and white gingham. Gingham is also good for a small girl's room, and you can use a small-patterned floral glazed chintz in a bathroom. You can also trim pine dresser shelves in this way to give them a softer look. If you want to conceal items on shelves, make slightly deeper curtains and suspend them from curtain track fixed to the shelf edge.

Cotton pillow lace, narrow broderie anglaise edging, a scalloped crochet valance (a traditional detail for shelf edging and window trim in canal barges), rick-rack braid, ribbon, curtain braids and trimmings including fringes and tassles, and simple swags of softly draped sheer fabric held in place with ribbon bows to hide the fixing tacks or staples will all add an elegant touch to the edge of bathroom or bedroom shelving.

Garlands of dried flowers, leaves, plaited rush edging, and other country-style trimmings will look equally effective in the right setting. Search the haberdashery, 'notions' and dressmaking departments of your local store, or look through baskets of old fabrics and remnants at car boot sales, and in charity or thrift shops for a bargain.

The shelves themselves can be given a soft touch – felt, green baize or almost any type of fabric (a plastic-coated, wipe-clean fabric makes sense for shelves which get lots of wear) can be stuck or pinned on, and the edge finished with decorative brass-headed tacks. Or you can make a 'sleeve' from fabric to totally enclose the shelf and then edge it with one of the trimmings suggested above. Use a fabric which is easy to wipe clean and which does not shrink too much – glazed chintz or a firmly woven cotton would be a good choice.

Self-adhesive plastic can be similarly used for an easy-clean surface, and can be neatly wrapped around the edges of the shelf. A border cut from the same material, or a wallpaper or wall vinyl border can be stuck or pinned to the shelf edge. Deckle edging or a diamond, scalloped or castellated effect can all add an extra design dimension.

For a more permanent fixed edge, pierced plywood or hardboard can be cut into a lacy patterned pelmet (cornice) with a fretsaw, and decorated to match or contrast with the shelving – paint these before fixing. For a simpler effect, cut a scalloped or zig-zag strip of plywood or hardboard, and paint it in bright colours before sticking or pinning to the shelf edge – a two-tone harlequin effect looks particularly effective in a playroom or family kitchen. Garden trellis can be stained or painted and similarly fixed to the shelf edge as a decorative trim. The wood should be sanded before decorating.

Some of these fixed edgings can be positioned so they project a little above the edge of the shelf to act as a retaining rail to prevent china and glass from slipping off the shelf. A rail or piece of beading or dowelling can be fixed just above the shelf to serve the same purpose.

Below: Maximum use is made of the space above and to the side of a small casement window. The shelves are softened with an embroidered fabric trimming normally used to edge linen.

DECORATIVE SHELF EDGING

△ **1** Measure the length of the shelf and decide on how many cut-outs will fit along it. As a guide to calculating, the triangles in the picture have a 50mm (2in) base and 35mm (1⅜in) height, giving an angle of 70 degrees. Make a drawing of a few triangles on paper to check the shape. Mark the total depth of the edging on the plywood (60mm/ 2⅜in in this case) and the baseline from which the top of the triangles will hang (25mm/1in from the top). Divide up the baseline into units equal in length to the base of a triangle. Midway between two of these divisions, on the bottom edge, make another mark from which to set the angle of the sliding bevel. Set the stock of the sliding bevel against the bottom edge and set the blade from this mark to one of the adjacent marks on the baseline. Clamp the blade in this position.

△ **2** Work along the baseline with the sliding bevel, marking down to the bottom edge from each mark in both directions, to form the triangular cutting line. Cross-hatch the waste portions which will be removed.

△ **3** Clamp the plywood to the work surface and cut out each triangle in turn, working from the bottom edge towards the baseline and cutting to the waste side of each line. Finally cut along the top edge. Smooth the cut edges with medium-, then fine-grade abrasive paper, rubbing towards the back surface (on which the lines were marked).

Left: *Geometric shelf edgings in bold colours help define the folk-art feel of this interior. The overall effect is achieved simply and inexpensively, but makes immediate impact.*

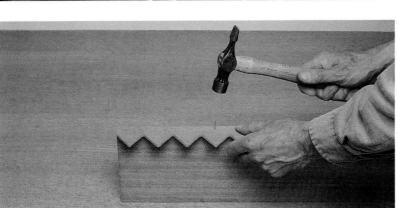

▲ 4 Hammer 15mm (⅝in) panel pins at 150mm (6in) intervals into the plywood, until the points just protrude, so that they will be central in the thickness of the shelf. Apply PVA woodworking adhesive (white glue) to the edge of the shelf, position the edging so that the ends and top edge are flush, and hammer in the pins. Punch the pin heads just below the surface of the plywood. Wipe off excess adhesive with a damp cloth. Mix to a paste some sawdust and PVA woodworking adhesive and use this to fill over the pins. Rub smooth and apply a finish as required.

CASTELLATED EDGING

Measure the length of the shelf and decide on how many cut-outs will fit along it. Make a drawing of a few castellations on paper to check the shape. As a guide to calculation, the cut-outs in the picture are 40mm (1⅝in) wide and 35mm (1⅜in) deep. Mark the total depth of the edging on the plywood (60mm/2⅜in in this case) and the baseline from which the top of the cut-outs will hang (25mm/1in from the top). Divide the baseline into units equal in length to the width of a cut-out. Remember in this case that you will need an odd number of units for there to be a protruding section at both ends. Set the stock of the try square against the bottom edge and draw a line up to each mark on the baseline. Cross-hatch the waste portions.

Clamp the plywood to the work surface and cut up to the baseline on the waste side of each line. To remove the waste, position the blade part way along one of the cuts and make a curved cut into the opposite corner. Then cut back along the baseline to remove the second corner.

CUPBOARDS AND CABINETS

If you have decided on fitted furniture (with or without shelving) to provide storage for some of the items in your home, you will still probably need the addition of a few freestanding pieces to provide a little extra character to your scheme, or serve as an extension or adjunct to the built-in units and shelving.

These will usually be such items as chests of drawers, cabinets, wardrobes and cupboards, which can be bought as new, ready-made items. Or you could again buy flat pack furniture and assemble it yourself; or you might prefer to seek out some bargains at local auction rooms and house sales, or in junk shops and from jumble (rummage) sales.

BUYING NEW

If you are buying new pieces you will obviously choose something you like; which will suit the style of the room, and perhaps add character and be in your price bracket. There is a wide choice available from many different furniture manufacturers to suit all tastes, so it is really a question of shopping around – don't buy the first thing you see.

Remember to take your room plan and measuring tape with you so you can check the dimensions to make sure the piece will fit into the space, and also measure inside to check it will hold everything you want to store comfortably. Also measure and check the door swing so you will be able to open doors fully, and take things out easily, when you get the item back home. Even if you are buying new furniture from a reputable shop, it is still a wise precaution to check it thoroughly – shake it to see if it wobbles. Take drawers out and check the runners and the bottoms of the drawers. Sometimes these can be so insubstantial they won't take the weight of more than a few items, so are no good to you if you plan to fill them with heavy books and papers.

Also push drawers to and fro to ensure they glide smoothly, and check handles or other fastenings to see they don't catch your fingers, or are likely to come off. If the handles are unusual, check on replacements or extras, and ease of refitting in case of breakage. Check hinges on cupboards, wardrobes, kitchen units or cabinets – opening and closing doors to make sure they are well hung. Look at any legs, or feet to make sure they are firm and well fixed, and that they don't stick out too far. The whole piece should be well balanced.

Above: *Open and closed storage cupboards, purpose-built above a tiled worktop in a colourful kitchen.*

If you are buying kitchen or bathroom cupboards and cabinets, there are many different styles from which to choose. Some of these are of rigid construction; others come packed flat for you to build; others are sold as part of a total installation package (the installer can even supply appliances) which will be installed on site. Make sure you know exactly what the supplier will do – are worktops included, and what about re-decoration of the room, tiling, and flooring?

Some ranges are made in solid wood, or MDF (medium density fibreboard) and others have glossy laminate finishes. Check that you are getting what you think you are: some units have wooden or wood-veneered doors, but the carcass is of inferior quality yet they are still called 'solid wood'. With laminate make sure there are no chips on the surface or around the edges.

MAKING DO

It is possible to renew doors and drawer fronts on existing kitchen cabinets. There are various ways of doing this, from paint or self-adhesive plastic to completely renewing the doors. You can sometimes do this yourself, if your units are standard size (*see* page 56), but there are also specialist door-replacement companies – look in local directories for names, addresses and telephone numbers.

Melamine surfaces can be painted if they are in good condition, and you can create interesting patterns on doors and drawer fronts to give kitchen units a new look.

If you want new work surfaces to tone, order these cut to size and shape (if necessary make a template) and stick in position on the existing worktop, using an impact adhesive.

PAINTING MELAMINE KITCHEN UNITS

▲ **1** Take off the door fronts and take out drawers. Scrub twice with hot soapy water and fine-grade wet-and-dry abrasive paper to remove dirt and grease. Leave to dry. Wipe over with lint-free cloth dampened with methylated spirit (wood alcohol). Paint the surfaces with a base coat of eggshell (gloss) finish – this will be the main colour so should be the paler one.

▲ **2** Plan the design to scale on a large sheet of paper. Cut a stiff cardboard template to size and, with ruler and pencil, gently mark the diamond pattern onto the prepared surface. Paint the diamonds in the second colour filling in the main blocks first and completing the edges with a fine brush. Leave to dry. If necessary the whole surface could be given several coats of clear varnish to protect. Refit doors and replace drawers.

BUYING SECOND-HAND

Old furniture – cupboards, cabinets and units – can also be given a new lease of life with paint, stain and various 'distressing' techniques; by stencilling; with the addition of mouldings and beading; with new handles or knobs; or by *découpage*. Some of the interesting finishes worth trying include marbling, stippling, staining-and-wiping, etc.

CHECK IT OUT

Of course you may find an attractive piece which only needs a little care to return it to its former glory, and which it could be a pity to paint, stain or disguise in any way. Check that it will not be too costly or difficult to restore. If it is expensive, always make sure it is what it is said to be – not made from several different pieces or artificially aged. The only way to be sure of this is to go to a reputable dealer, or to ask for advice from a professional.

Before you buy second-hand furniture perform all the checks that you would do for any item of furniture. Shake it to ensure it will not wobble in use, and is firmly constructed. If not, it may be able to be strengthened with metal or wooden struts or corner pieces. Check any feet or rails for firmness and for any sign of rot – if it has been standing on a damp floor for some time this may have 'eaten' into the base.

Take the drawers out of chests of drawers, units etc and check the runners for firmness and ease of opening. If they stick it could be that the drawers have been put in the wrong places, so try changing them around. If the runners are worn they can be replaced, but don't combine new plastic runners with old furniture – wood is best. If really necessary, you can use metal runners.

Look at the base of each drawer. Make sure it is not split, and that it is firm enough

to take the weight of the items you want to store. Again it may be possible to strengthen the bottoms of drawers, or to replace them completely.

Check any hinges, handles and catches to make sure they work – and also for authenticity. Some old knobs or handles may have been removed and replaced with ugly new ones. Of course you can totally alter the look of a piece with new handles. Hinges may also be wrong, or ugly or awkward, but again these can be replaced fairly easily. Many specialist companies produce these fittings.

Don't forget to check for woodworm. The sign of this is small holes, rather like those on top of a pepperpot, where the grubs have bored their way out of the wood – and there may be fine wood dust on the floor under the piece. If there is bad infestation, don't buy If it is only mild, then you can risk it as long as you treat the piece with a suitable product before you bring it into the house. It is essential to treat it before the furniture is in contact with other wood in your home, since woodworm can attack not only the furniture but the whole structure of the building.

Left: This attractive aged look is simple to achieve. Paint on an undercoat of green emulsion (latex). When dry, cover with a top coat of blue. While the blue emulsion (latex) is still wet, gently wipe with a dry cloth to reveal the green paint underneath.

Above: Embellish wooden furniture with stencilled designs. Make the stencil from clear acetate or stiff card (cardboard), using a craft knife to cut out the pattern. Tape the stencil in position and apply paint using a flat stencil brush.

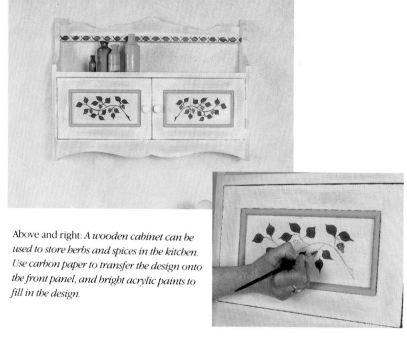

Above and right: A wooden cabinet can be used to store herbs and spices in the kitchen. Use carbon paper to transfer the design onto the front panel, and bright acrylic paints to fill in the design.

And don't forget to take the measurements of both the outside (width, depth and height) and the inside of the piece, so you can be sure it will fit exactly where you want it to; that it will store the items you intend to put in it; and that you can actually get it into your home, up the stairs and along corridors.

BACK TO BASICS

If you intend to paint, stain or use any other decorative finish, this will probably mean stripping back to the bare wood, preparing the surface and starting again from scratch. There are various ways of tackling this.

If you are buying an old piece, or thinking about painting one, remember that if the furniture is veneered it cannot be stripped. Any form of caustic stripper will also dissolve the glue, causing the veneer as well as the paint to float off. It would be a pity to paint over a beautiful veneered piece anyway, but there is no reason why you should not paint a tatty veneered item. But you will have to repair any chips in the surface, or replace any missing pieces of veneer, to give you a smooth surface to work on.

To strip a solid previously painted piece you might take it to a firm that strips doors and pieces of furniture in a bath of hot caustic soda – this saves a lot of time and work. The treatment removes all traces of paint from intricate mouldings and curved surfaces, but it is possible for the hot caustic bath to dissolve glue – so you may be given a pile of pieces of timber when you go to collect your furniture. Doors sometimes warp, too, so discuss this with the expert on site before making a decision.

It may not be necessary to strip a well-painted piece of furniture at all – you can just wash it down with soap or other proprietary cleaner and water; wipe it dry and then

sand-paper lightly to provide a slightly roughened surface as a 'key' to take other coats of paint. It is always wise to use a basic undercoat which is recommended for use with the topcoat. It should be suitable for use on woodwork – don't use emulsion (latex) paint on furniture. Rub down lightly between coats to get a perfect final surface.

STRIPPING TIPS

If the paint is in a bad condition and you have decided not to risk dip-and-strip, you will find that using a chemical stripper is the best method for furniture. Hot treatments such as blowtorches and hot air 'guns' can easily scorch, or even burn old wood.

There are two kinds of chemical stripper: liquid and paste. Both are very caustic and must be used with care. Try to do the work outside, or in a well-ventilated garage; protect the surrounding floor with old newspapers. Wear safety goggles or glasses and protect your hands with rubber gloves. If you have an accidental spill, wipe it up immediately. Rinse splashed skin or clothes with plenty of cold water.

For liquid strippers use a clean old paintbrush to apply the clear liquid; the paint should wrinkle and start to break up after about 15 minutes, but follow the instructions carefully. Give it enough time to work so you don't have to re-apply the stripper, but don't let it dry or the paint will begin to harden again. With some types it is recommended to cover the surface with plastic or paper to prevent drying. Use a scraper on large flat

Above: *Give a storage chest a new lease of life. Mark out the drawer design in chalk and paint in acrylic paint.*

surfaces, pushing it away from you. Use a shavehook on small surfaces and intricate mouldings, pulling the tool towards you. Clear up all the perished paint immediately. With a heavy build-up of paint you may need to use a second application.

With paste strippers, apply a thick coat, again following instructions, with a clean old brush. It is usually recommended to cover the paste with plastic while it activates, and to spray occasionally with water. After the recommended time (usually hours rather than minutes), scrape away the paste as above, bringing the old paint away with it.

It is essential to neutralize stripped wood before redecorating – the recommended product and method should be mentioned in the instructions with the stripper.

You will then need to rinse the piece of furniture (don't make it too wet) and let it dry. This is then the moment to attend to any repairs, and to fill any cracks. Rub the filler down to a smooth flat surface before redecorating or staining.

REFURBISHING

You can, of course, restore and refurbish old furniture – or give some of your existing items a new lease of life.

There are some exciting design possibilities including various *trompe-l'oeil* treatments such as marbling. Some old panelled wardrobes and cupboards can look amazingly good if you replace the panels with glass – stained, decorative, sand-blasted or plain. Sometimes you don't want items in glass-fronted cupboards and cabinets to be on view, in which case lining them with pleated fabric against the glass can provide privacy and instant co-ordination with the room scheme. In some rooms, wire mesh, gauze or chicken wire is an interesting replacement for panels or glass.

When you are refurbishing old pieces don't forget the inside. Many old chests of drawers for example, have a rather rough finish inside the drawers, or cupboard interiors can look stained and worn. You can paint the inside of some pieces in a matt or gloss paint, but there are other ways of lining. Use fabric or green baize (the traditional lining for cutlery drawers), but don't stick it down – fix with drawing pins (thumb tacks) so it can be removed for cleaning. Attractive lining or wallpaper provides an easy-to-change lining and silver or gold foil can look effective in brightly coloured drawers.

PAINTING AND STAINING

Once cabinets, cupboards, chests and other items of furniture are in a good state of repair they can be decorated, and trimmed in any number of different ways (*see illustrations on pages 50–55*). Painting is a quick way of creating a new look and introducing style and colour into a scheme. It can also help you to integrate another piece into a room or run of storage successfully, so that it co-ordinates with existing furnishings and the items, however disparate, look all of a piece.

You can use a plain colour, or do some exciting things with painted panels, or patterns and pictures in different colours; or use any one of a number of popular painting techniques such as marbling, stippling, graining, dragging or ragging. Another idea is to decorate the piece with pre-cut stencils – or make your own to echo a particular design feature, or create a specific look. Most of these techniques will require a smooth, dry pre-painted surface, although you can stencil onto plain or stained wood. Some very effective results can be achieved by using a contrasting stain rather than paint for the stencilled motif.

Above: *Rich red and golden stencils transform this plain pine chest, simply and inexpensively.*

DESIGN TIP

The easiest way of dealing with a piece of furniture to be decorated is to take it outside, or into a well ventilated garage or shed. Stand it up, or lay it down on a trestle table so you can reach it easily and walk around it. Take the drawers out (number them underneath in pencil or chalk so you remember the order for replacing them) and, if it is an easy job, remove doors and hinges. It is much quicker and easier to paint a door flat, and drawers separately from the carcass, and you will also avoid any unsightly 'sags and runs' on the finish.

PAINT

Always use the correct type of paint recommended for woodwork (usually an oil or solvent-based gloss or semi-gloss finish, although there are now water-based gloss paints available). If the wood is raw or porous, or you have used a chemical stripper on it, apply a suitable wood primer first. Also treat any knots with a proprietary knotting to prevent seepage of resin, which will break through the new surface and spoil it. It is usual to apply one or two coats of undercoat

first, lightly rubbing down between the coats, and then the final topcoat – but the more coats, and rubbing down between them you do, the better the finish will be.

STAIN

There are many different types of stain, mostly in various 'wood' tones, for both indoor and outdoor use – the latter are usually intended for rough-hewn wood, and may not be suitable for furniture. Never use coloured creosote for indoor woodwork, as it gives off toxic fumes. Some stains require a primer on bare wood; others act as their own primer, but need several coats; others will darken considerably the more coats you apply; some give a glossy finish which builds up the more coats you use.

Other stains are matt (flat) and sink into wood like ink into blotting paper. These require a finishing coat of polyurethane or other varnish to seal them – this can be a matt seal, a semi-sheen, a gloss or a high gloss; the choice is up to you, but a very glossy finish on a stained piece can sometimes have a 'sticky' look.

Stains also come in some exciting colours, from bright primaries to soft and subtle shades. Because they are transparent, the grain of the wood will always show through – which is part of the charm of staining.

Stains are usually selected from a small colour card, or even from the label of the container. It is essential to test the effect before you apply the stain. Many of the

wood-effect finishes in particular can come up looking very dark, and nothing like you expected – 'dark oak' may well be almost black – and the original colour of the wood will also influence the finished result.

Some companies provide little 'testers' so you can buy a small quantity and test it on a similar piece of wood to the furniture, or on the underside or back of the piece, which won't be on view (but take care, since this might not be the same wood as the rest of the carcass). If a tester is not available, buy the smallest quantity of stain you can and do a test first before buying a larger quantity.

Follow the manufacturer's instructions for applying the stain – this can be brushed or rubbed on, or a bunched-up cloth may be recommended. Stains, unlike paint, don't usually come in aerosol cans, so can't be sprayed on. Ensure the coverage is even, then leave the stain to dry for the recommended time before applying a second or subsequent coat, or the suggested sealer.

For a softer look you can use a stain-and-wipe technique – apply the stain with a brush, leave for a few seconds and then wipe off with a soft cloth. This technique looks particularly effective on wood with an attractive grain, and you can use several colours to build up an 'aged' look – or put a pale tone over a dark one to give the furniture a slight bloom. You can also use a wipe-on-and-off coloured stain to create a 'distressed' look on the edges and corners of larger items of furniture.

DECORATING AN UNFINISHED PINE CHEST

▲ **1** Old or new solid wood can be painted but take care with veneered pieces as the veneer might start to peel when over-painted, and it cannot be stripped and re-painted using caustic stripper.

▲ **3** Cut a template for the star shape and establish correct position (centre on the handle holes). Draw round this with white pencil or chalk.

▲ **4** Paint stars and handles in contrasting colour (gold was used here). Replace handles.

▲ **2** Remove handles, catches, and knobs. Wipe down the piece to remove any dirt or grease. If necessary wash with soapy water and fine-grade wet-and-dry paper. Leave to dry thoroughly. Wipe with a lint-free cloth dampened with methylated spirit (wood alcohol). Paint with primer (if necessary) and undercoat. Paint the piece in the desired colour using a gloss finish. Do the carcass first – it is easier to paint drawer fronts and doors if you take the drawers out and the doors off their hinges. You can use a brush or a small roller to do the painting.

▲ **5** If necessary the chest can be given a couple of coats of clear varnish to seal and protect the paint.

REFURBISHING CUPBOARD DOORS

Left: *Tired kitchen units can be transformed quickly and inexpensively with a special peel-and-stick decorative laminate for kitchen units which comes in a pack. Each contains enough to do two base unit door and drawer fronts, or one large unit.*

▲ **1** Ideally remove the door from the cabinet carcass, take out the hinges and lay it on a firm, flat surface. Peel back 25mm (1in) of backing paper and position top edge so it butts up to door moulding (or slightly overlapping square-edged door) and overlaps edges sufficiently to cover them. Pull away backing paper with one hand while smoothing down the decorative sheet with a soft cloth from the centre outwards. If any bubbling appears, peel up to the bubble and re-smooth. Continue to just beyond the bottom edge of the door.

▲ **2** With the blade of a trimming knife held flat against the adjacent face of the door, cut off the surplus covering flush with the edge. Where the covering is turned onto the door edges, make angled cuts as as necessary so that there is no overlap.

▲ **3** With a sheet of fine glasspaper (sand paper) wrapped round a cork sanding block, and held at an angle of 45° to the corner, take off the rough edges left by the knife. *Warning:* The glasspaper will scratch adjoining surfaces, so take care when sanding.

REPLACING CUPBOARD DOORS

▲ **1** Many kitchen cabinet hinges are mounted on a baseplate screwed to the carcass. To remove the door, one screw per hinge is slackened to allow the hinge to be slid from the baseplate. A second screw is for adjustment purposes only. After removing the door from the carcass, remove the screws holding the hinge bosses in the door and take out the hinges.

▲ **2** Holding the old door on top of the new one, flush with the top edge, mark the distance of both hinge recesses from the top edge. Repeat with the side edges flush to mark the distance in from the side.

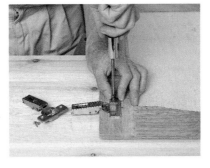

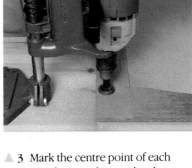

▲ **3** Mark the centre point of each recess on the new door and make an indent with a centre punch (or sharp nail). Clamp the door to a work surface. Wearing safety goggles, cut out the recess to the same depth as the original one with an end mill and electric drill (ideally using a drill stand).

◄ **4** Fit each hinge boss into its recess, with the hinge arm square to the edge of the door. Mark through the fixing holes with a bradawl or gimlet and screw the hinge into the recess. Replace the door on the cabinet and tighten the fixing screws. Adjust the position of the door with the adjustment screws, if necessary, until it sits squarely and closes flush to the face of the cabinet.

REPLACING DRAWER FRONTS

▲ **3** Using the screws for positioning, hold the new false front against the drawer and screw in place.

▲ **1** Unscrew existing false drawer front from inside drawer.

▲ **2** Clamp old drawer front over new one, all edges flush, and drill pilot holes through the existing fixing holes. Use a depth stop, or insulating tape wrapped round the drill bit, to ensure that you don't drill right through the new front.

4 If you don't want to damage the old front, remove it and position the new front on the drawer. Mark through the old fixing holes, remove new front from drawer, drill pilot holes and refix with screws.

CHICKEN-WIRE DOOR PANELS

▲ **1** The door panel sits in a recess in the edge of the surrounding frame and should not be glued in. To remove the panel, drill starting holes for a jigsaw blade and make two cuts along the length of the panel and one across the middle. Lift up the edge of one centre strip and, with a little force, pull it away from the recess in the frame. Repeat for the other half. Remove the middle portions of the remaining panel in the same way. To remove the corner pieces, drill a hole through to take the blade of an old screwdriver. Stand the door on edge, insert the screwdriver through the hole and tap it sharply with a hammer. If this method isn't successful, the only answer is to cut round inside the frame with a jigsaw and sand smooth.

▲ **2** Using 15mm (⅝in) hockey-stick moulding as a guide, mark on the back of the door frame where the edge of the netting will come to. Align a finished edge with one of the guidelines and adjust the netting to reach the other lines, so as to leave a closed mesh at each edge, if possible. If the mesh is not square to the edges, pull at diagonally opposite corners to correct it. Temporarily staple the netting to the frame.

▲ **3** Cut off the surplus netting with tin-snips, pulling the cut portion aside as you proceed to prevent scratching. A useful hint is to slide one end of a length of thick rubber tubing on to each handle of the tin-snips to make them spring open between cuts.

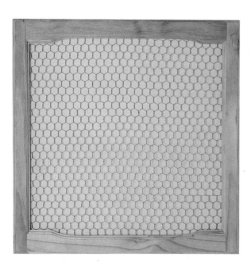

Above: *Chicken-wire panels strike a country note in the kitchen and reveal the contents of the cupboard while providing an element of protection.*

▲ **4** Cut hockey-stick moulding to length in a mitre block so that the rebate overlaps the edge of the netting and the rounded edge rests against the door frame. Hammer in the staples fully and pin the moulding to the door frame. A little woodworking adhesive on the mitred ends of the moulding will give added strength.

PUNCHED-TIN DOOR PANELS

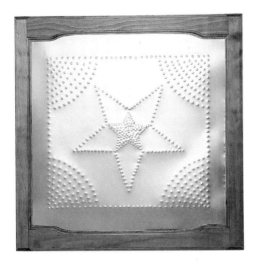

Above: *Punched-tin panels evoke the American folk art style. They can be simply fixed on top of ordinary cupboard doors.*

▲ **1** Make a template of the door panel by laying a sheet of brown paper over it and pressing it into the edge of the door frame all round the panel. Cut it out and lay it on the tin sheet; mark round it with a felt-tip pen.

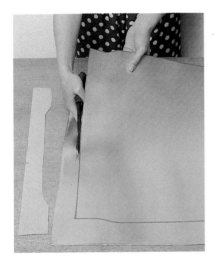

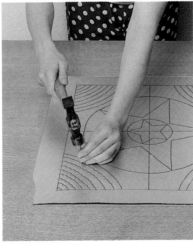

▲ **2** Cut the tin to the panel shape with tin-snips and check for fit against the door. Adjust as necessary until there is no bowing of the tin. Flatten any undulations at the edges by hammering gently on a hard surface.

▲ **3** Mark out the area to be punched and draw centre lines in both directions to ensure that the pattern is symmetrical. Plan out the pattern on the paper template, then transfer it to the tin. To punch the pattern, work systematically along the lines, punching at regular intervals and with the same number of hammer blows, to achieve a uniform depth of indentation. Where the pattern comes to an angle, make an identation at the point and work away from it in both directions.

▲ **4** To stick the tin to the door panel, coat both the tin and the panel with impact adhesive and allow to dry as specified by the manufacturer. Ensure good ventilation. Place one edge of the tin carefully in position on the panel (the bond is almost instantaneous) and prop up the opposite edge on offcuts of wood. Working from the centre of the fitted edge outwards (to avoid trapping any air beneath it), press the tin onto the panel with a soft cloth. Gradually work your way across the tin in this way towards the propped edge, then finally remove the props and press down the final edge.

SITING STORAGE

When you are measuring up and planning where to put cupboards, shelves, units, etc try to make the most of any 'natural' space which already exists, so the storage is fairly unobtrusive, and does not project too far into the room.

Storage cupboards or shelving can be tucked into an alcove, and window seats, low units or shelves may be positioned under bay or dormer windows. Small 'awkward' corners between a window and return wall can be filled with shelves or slim cupboards, but the curtain treatment will have to be tailored to fit into the reveal and not obscure the storage. Triangular spaces under sloping eaves or under the stairs can often be a good place to put a storage unit, desk, chest or dressing table with shelves above.

Of course, in some cases you can create a complete storage wall, which may give a much-needed focal point to a living-room –

a square box-like area with flat walls can be made to look much more interesting with shelves and units running right across from wall to wall, with space left in the centre for a gas or electric fire – or somewhere to place the television set.

In a bedroom, an alcove can be created for the bedhead by leaving a gap between wardrobe cupboards to take the bed, and possibly bedside tables, and the whole linked with top cupboards above. With a single bed, the side of the bed could be placed parallel to the wall, with cupboards built at the head and foot linked with the top cupboards above, and curtains fixed to close across at night making a cosy 'bed-in-a-cupboard' – very popular with children and teenagers. This same treatment can be tried in a larger-than-average size bathroom too, with cupboards to each side of the bath and top cupboards above, forming an attractive recess.

Left: *The maximum amount of storage space is created in this country living-room – units and shelves are built into the fireplace recesses, and window seats with cupboards are built under a bay window.*

MAKING THE MOST OF NATURAL SPACE

You will find many ideas throughout this section for making the most of available space – adding storage drawers under beds and other pieces of furniture; fixing shelves and racks inside cupboards, into 'dead' corners and behind doors. But don't forget the space above your head! In tall rooms, storage shelves can be built above doors or cupboards can be mounted at the top of stair wells (make sure they are high enough up and firmly fixed to prevent accidents). In a bedroom or bathroom, slim cupboards can sometimes be fixed above the bed or bath, screwed firmly into the ceiling (again make sure the fixing really is firm) and used to suspend curtains to simulate a four-poster bed.

Often the most sensible place to install storage and shelving is in a natural recess or alcove – many houses have a projecting chimney breast in the living room and bedrooms, with recesses to each side, which can be used for the shelf supports, or to take fixings for slide-out drawers and wire baskets. These recesses can be converted into wardrobe cupboards by the addition of doors, curtains or blinds – an inexpensive way of converting available space into workable storage.

Recesses are a good place to install wardrobes too, but you will need to do some careful measuring – a narrow recess may not provide enough depth to hang garments sideways. It sometimes makes sense to extend the cupboards forward into the room, so they are deep enough – and if the fireplace is not used, they might be linked across the chimney breast with narrow shelves. The alternative may be to plan hanging space flat into the alcove rather than sideways, creating a slimline cupboard.

Sometimes there is a problem with the window, if the chimney breast wall is at right-angles to the window wall: a cupboard may fit into the one recess, but in the other a cupboard may block some of the light, and come partway down the window and be visible from the outside. There are several ways of coping with this. You could chamfer the cupboard, so it is neatly angled into the window or you could build a low unit the width of the window, which extends up to the window sill height. Add narrow shelves, or a slimmer cupboard above this unit. Consider building a window seat under the window, put a cupboard into the space on the other side of the window, and fix a stable door on the cupboard in the recess, to make access easier. There are numerous permutations on the theme, which will not only provide much needed space, but will add character.

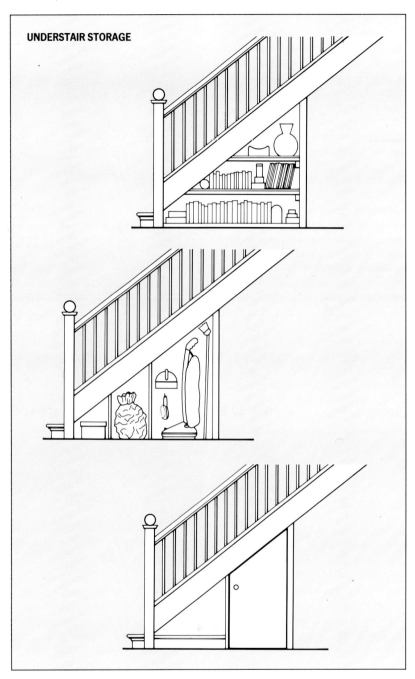

UNDERSTAIR STORAGE

Above: *There is often a great deal of space underneath the stairs, which is not used effectively. Adjustable shelves may be used to store books, records etc, or cleaning utensils and tools. To hide unsightly items, the area may be panelled and fitted with a door.*

UNDER STAIRS AND EAVES

The triangular space under a flight of stairs is often closed in to form a cupboard, which in many households is totally disorganized and filled with clutter. If you want to keep it as an enclosed space, there are many practical ways of using the area.

If the space is large enough to be converted to a walk-in cupboard, you can install wall-mounted shelves all round the three sides, and below the slant of the stairs themselves. It is often practical to have a wide shelf about 1m (3ft) from the floor, with wire pull-out baskets or plastic stacking crates below, and narrower shelves above. It is even more practical if these are adjustable. It is a good idea to store little-used items at the back and on the higher shelves.

Special hooks can be fixed to the walls for some items; screw-in hooks can be placed on the edges or undersides of shelves; small things can be stored in screw-top jars with the lids fixed to the underside of a shelf; the back of the door can accommodate hanging 'pockets' made from plastic or fabric; and you can install extra narrow shelves, or special holders for the iron, ironing and sleeve boards, etc. Don't forget to light the inside of such cupboards or you may find they are not as practical as you had hoped.

If you need to accommodate certain large items, such as golf clubs and other sports equipment, tools, the vacuum cleaner or a push-chair (baby buggy), make adequate provision for these. Measure them first and work out how best to store them – vertical racks can be useful for rackets, hockey and lacrosse sticks for example. Notched shelves can hold broom and mop heads, or the hose and nozzles for the vacuum cleaner. These are also useful for holding gardening tools, whether outdoors, in the basement, or under the stairs.

The very narrow triangular corner under the lower stairs is often impossible to reach from inside the cupboard, and so the space is wasted. Consider installing a separate small door to give easier access, and fit out the inside with pull-out baskets, toy boxes, or drawers on castors or wheels. Or it can be left open and fitted with cardboard or plastic rolls or short lengths of drainpipe which are easily stacked in a pyramid shape, and can be used to store wine bottles, or boots and shoes.

OPEN UP!

If it is possible to store things elsewhere in the house, you can open up the space under the stairs to provide a practical extra 'room' with its own storage. It may be an ideal place to place an under-stair shower – at the highest point of a deep staircase. This would be a particularly practical proposition in a three-storey house, with the shower fitted into the first-floor landing.

Showers are now easy to install thanks to thermostatically controlled, instant water heaters, with a booster pump that feeds the water to the shower, so you don't need to worry about positioning a water tank above the shower. The remaining triangular space to the side could be fitted out with dressing table and basin, or used for linen storage (upstairs); or a hobbies area or place for toys and games (downstairs).

Above: This under-stairs area is sufficiently large and open to comfortably hold a graduated shelving unit. The short lengths of shelving easily hold the weight of the books, and the units can be rearranged as necessary to suit other locations.

Under the stairs can also be the ideal place to site a small home 'office' – two chests or filing cabinets placed to each side and linked with a desktop, or a suitable-sized desk. Wall-mounted shelves above can provide space for files and papers, reference books, the telephone or a fax machine. There may also be space for a computer or the family 'noticeboard'. In some homes this would be an ideal place to settle a child to do homework while the evening meal is being prepared in the adjacent kitchen.

In some homes, the stairs actually form part of the living area and rise up through the main room to the floor above. In this case the space under the stairs may be the ideal position for a folding table flanked by two chairs; or to wall-hang some units combined with a moveable serving trolley. Or it could be the ideal place to build a unit to store bottles, glasses, and items needed for the dining table. The top can be used to pour drinks, or as a food serving area. A triangular wine rack can be used to hold bottles.

Above: *Custom-built shelves and wardrobe use this area under the eaves to best advantage. The tiny area to the right of the wardrobe includes shelves, too, ideal for personal items. The blind (shade) is perfect for protecting the contents of the wardrobe, and doesn't take up any additional space when open.*

USING ALCOVES

If you use an alcove for storage you will want to hide the clutter from view. One way of doing this is to use a roller or Venetian blind (shade) (*see* pages 66–67) fixed to a pelmet or batten across the top of the alcove. This only works well in a fairly narrow alcove, since a blind only operates efficiently if the span is not too great and not too much strain is put on the mechanism. Manufacturers give recommended maximum widths in their measuring-up instructions.

Other types of blind (shade) can also be used – louvred blinds or shutters made from wood or plastic can give a colonial look to a room. Rattan or cane strip blinds, pleated blinds, even ruched or tailored Roman blinds will all work well if the material is chosen with care. It must be strong and dustproof – choose a firmly woven textile such as cotton, a linen union (mixed with other fibres) or a glazed chintz for best results.

Vertical louvred blinds (shades) can also be used across alcoves. When they are open, the slats concertina back neatly into the 'parking position, and take up very little space – so permitting full access to the alcove. They can be fixed to the ceiling, or hung from a pelmet or batten across the alcove.

In some settings blinds may look a little harsh – in this case consider using curtains. These can be made from the same fabric as the window curtains, and lined (and interlined) if necessary so they hang well. In a narrow alcove one curtain will suffice, pulling to one side for access. Where there are alcoves to each side of a chimney breast, one curtain should pull to the left and the other to the right for symmetry. The track or pole can be ceiling-mounted, or fixed to a pelmet or batten.

In a wider alcove, a pair of curtains may be necessary, dividing down the centre. Here a fixed heading may be preferable, with tie-backs or curtain bosses used to take the weight of the fabric when access to the alcove is required. Alternatively, a track with lap-over at the centre can be used; this will have to be mounted on a firm pelmet board.

A panel of heavyweight fabric might be more suitable for some rooms. Sailmakers' or awning canvas can be bought from yachting specialists and ships' chandlers. It can be cut and seamed to create a bold geometric pattern, or painted with fabric paint. It could be fixed to a batten or hung on a curtain pole with tailored loops or ties. A batten at the bottom will hold it neatly in place.

Above: *Heavy canvas with bold geometric panels closes off an alcove. The fabric panels are both practical and a striking design feature.*

Above: *An otherwise unused corner is converted into a stylish shelving area.*

FITTING DOORS ACROSS AN ALCOVE

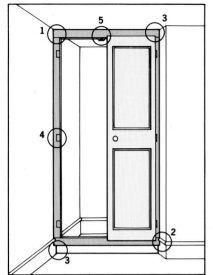

Left: *An alcove to the side of a chimney recess or in an awkward corner can be converted to a cupboard by installing doors. If the alcove is deep enough for coat hangers, install a dress rail to make additional hanging space for clothes. A shallow recess could be used for shelving, or for storing cleaning utensils suspended on hooks.*

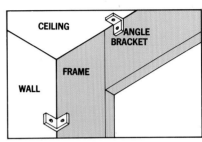

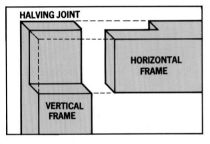

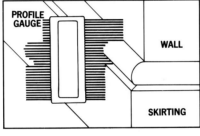

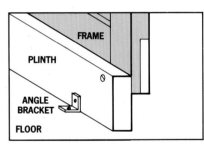

▲ **1** To fit doors across an alcove to convert it into a cupboard make a frame using 25 × 75mm (1 × 3in) timber. Join the corners together with halving joints (*see* page 66).

▲ **2** Hold the frame in position, and plane the edges to follow the outline of the walls for a neat fit. Use a profile gauge to copy the shape of the skirting (base) board onto the edge of the frame and cut out.

▲ Fit the frame in position with angle brackets, where it meets the floor and ceiling.

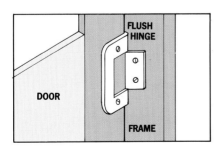

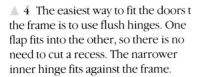

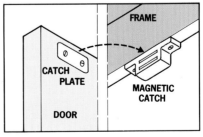

▲ **4** The easiest way to fit the doors t the frame is to use flush hinges. One flap fits into the other, so there is no need to cut a recess. The narrower inner hinge fits against the frame.

▲ **5** Fit a magnetic catch by screwing the main part to the frame. Place the metal catch plate over the magnet and close the door firmly to mark the plate position. Screw the plate to the door.

MAKING HALVING JOINTS

▲ **1** Mark a line across both pieces at a distance of the width of the timber from the end. Mark half-way down both sides from the end of the line.

▲ **2** Set a marking gauge to approximately half the thickness of the timber and make a mark from both faces of the timber. The points should coincide. Adjust the gauge as necessary, and mark from the pencil lines to the edge of the timber and across the end, and cross-hatch the portion of timber to be removed.

▲ **3** Cut from the face of the timber down to the thickness line and then from the end down to this cut. The two pieces should fit closely together. Join with PVA woodworking adhesive (white glue) and clamp, checking the angle with a try square. Leave until the adhesive has hardened.

DRESS RAIL IN AN ALCOVE

Left: *An alcove in a bedroom can be fitted with a shelf and dress rail to provide additional storage space for clothes. A blind (shade) hung from the shelf pelmet conveniently hides items away from sight. For instructions on putting up shelving in a recess, see pages 38–9.*

◁ **1** The end plates can be mounted on wooden blocks on a masonry wall to save having to drill closely-spaced holes for wall fixings. A central hole with a 50mm (2in) No.10 screw should be adequate. Position the plates at half the depth of the alcove, less the thickness of any door. In shallow alcoves the clothes can be hung at an angle to the rail. Mark through the fixing holes on the first plate, and fix in position. ▷

▲ **2** Wrap masking tape around the rail as a guide and cut to the required length with a hacksaw so that it fits between the blocks with about 2mm (⅛in) clearance. Remove the rough edge with a half-round file. Slide the second end plate on to the rail, with the flat face outwards, position the free end of the rail in the fixed plate, and screw the second plate to the block. For long rails, an intermediate support will be needed which is screwed to the underside of the shelf. A spacing block may be required to bring it down to the height of the rail.

▲ **3** To fit a blind (shade), screw angle brackets to the shelf pelmet, about 60mm (2½in) from the end.

▲ **4** Screw the blind (shade) support brackets in place. This particular blind had plastic inserts which push into the end brackets – one for the control end and one for the pin end.

◀ **5** Position the pelmet (cornice) against the shelf, and screw the angle brackets to the underside of the shelf. Hang blind (shade) in position.

EXCAVATE AN OPENING

Some kitchens have chimney breasts too, where the cooking range used to sit – if you don't intend to use the opening for a modern version or a gas or electric cooker, this is an ideal place to build shelves for display and storage (close up the chimney 'throat' first and make sure the top of the chimney is closed up). Decorate the back of the resulting alcove to contrast with the rest of the scheme, so it becomes a focal point; illuminate with concealed lighting inside the breast, or strip lighting under a pelmet or baffle, or accent with spots. If you prefer, the opening could be fitted with doors to make a cupboard (see page 65). An attractive treatment to consider is tiling the back of the alcove, and using glass shelves and doors so the tiles are permanently on view.

Right: Alcoves of any size can be used for storage. The narrow area in this living-room is fitted with shelving for display purposes.

A similar idea can work well in a sitting or dining room – especially in a 'through' room which has been made by knocking two small rooms into one larger area, and where one fireplace is now redundant. The fireplace opening can be used to house the television set, music centre, video, or to display a special collection. With the addition of doors, it could be used to store bottles and glasses; records, CDs, video cassettes and tapes; items needed for the dining table; or files and papers. This is also a possible treatment for a chimney breast in a bedroom or playroom. The opening could be used for additional storage or it could be used to make a dolls' house, castle or garage if the doors are painted with a suitable façade.

SPACIOUS SHAPE-UP

The interesting triangular shapes created by sloping roof spaces, areas under the eaves or the space under the stairs can present an unusual challenge but is often an area which provides much needed space.

In a bedroom, or bathroom under the eaves, the recess can be used to take a chest or dressing table, or possibly to build in the wash basin. Add mirror and shelves above to take cosmetics and the result is a neat make-up corner. In a studio apartment or teenager's bedroom, the space could be used to take a desk, or a unit for a computer/VDU/ television set, with shelves above.

The understair area can also be the ideal place to site a small study corner, with a desk, or two storage units joined by a work top. Add adjustable shelves above, which will fit neatly into the triangular shape, plus a swivel or folding chair and you will have a place to do the accounts, or persuade children to get down to homework where you are near at hand.

This is also a good place to store wine (lots of wine racks already come in a triangular shape). You could make a drinks corner, with low cupboards to hold glasses, bottles etc and a practical serving top which could be made from marble, laminate or a resin-based product – or tiled. Build in wine racks above, or shelves and narrow cupboards to provide more storage space for glasses. This space can also be used to store items for the dining table – particularly practical if it is sited between the kitchen and dining area.

Above: *A blocked-up doorway is used for book storage. When the door was removed, the wall was made flush on the hall side, the architrave was retained on the room side and the resulting recess was filled with shelves.*

Above: *Even awkward spaces can be used advantageously – shelves in a fireplace recess allow for the sloping ceiling.*

ESSENTIAL STORAGE

The ubiquitous television set and video recorder have replaced the fireplace in many homes as the centre of activity – the main feature of the room around which the family clusters on winter evenings. And that's far from the only piece of electronic equipment. A computer with a printer, a fax machine, and a telephone answering machine may also have to be fitted into the main living area.

Even the most ardent addict would agree that such equipment, not to mention the stereo system, and all the associated CDs, records and tapes, is not the most visually attractive. These items are dust-collecting and vulnerable to damage and breakage, and are better stored away out of sight. Much the most practical way of dealing with the problem is with custom-built storage. In many homes there are recesses to both sides of the fireplace in the main living room, which provide a natural place to install low cupboards for the bulkier items, with narrower shelves or cupboards above. In a dual-purpose living and dining room, one of the units can be used to store items for the dining table.

Where there is a blank wall, the storage can fill the whole area, and be designed to provide a focal point or add extra visual interest to the room. Sometimes the area around a casement window is a more practical place to site such storage. In a bedroom or bedsitting room, designating part of a run of wardrobes for such items may be a sensible option – as long as the television screen can be easily viewed from the bed.

Whichever solution you choose, storage should be designed to blend in with the style of the room. It is often better to treat it to match the walls rather than the other items of furniture in the room – this gives greater flexibility for future furniture arrangements and furnishing schemes. If built-ins are impractical because you are always on the move, or live in rented property, there are many free-standing storage systems available in a variety of finishes and woods, which are tailor-made for such equipment. Again consider textural contrast, style and form when choosing, and try not to end up with everything so carefully matched that your home resembles a furniture showroom. Although many units are designed to match one another, you may prefer mixing pieces that suit your taste and budget.

Above: *The television set is concealed in a cupboard in a living-room recess, and mounted on a pull-out shelf which can be angled for viewing.*

Left: *Freestanding units are adapted to take the video and music centre and also to store books, cassettes and video tapes.*

ENTERTAINMENT SYSTEMS

The television set, hi-fi and video recorder can all be put inside the cupboards. If necessary, a pull-out shelf, sliding section or drop-down flap, or a turntable can be incorporated to angle the TV set towards the main seating in the room (or the bed) for ease of viewing. Storage for records, CDs and tapes can also be made to pull out.

Shelves of the correct dimension are usually the most sensible way of arranging storage inside the cupboards, but in some cases drawers can be used, becoming part of the design feature of the units; or wire racks or internal drawers might be a better choice. Use racks or dividers to keep records and similar items neat and upright, so a few can be removed without the entire contents toppling over.

In some more streamlined or hi-tech interiors, a portable trolley may be the preferred option for such storage; or glass shelves can be built in to house them – but careful cleaning and dusting is essential to keep everything running smoothly.

Above: *A neat, low unit houses the television set and music centre on top and has drawers and cupboards below to store all the accessories.*

UNDER-BED STORAGE

There is often a lot of wasted space in a home, both above your head and under your feet, which can be used successfully for storage. In particular, the space under the bed is often wasted. Divan beds these days often come with built-in storage drawers. Others look as though they have a solid base, but in fact it is hollow, and can be reached by taking off the mattress and lifting up the top, rather like opening a padded ottoman. This type is only practical for long-term storage, since you don't want to have to be continually removing the mattress.

If you are buying a new bed look at the options. Some have full-length drawers, others two half-length ones; and double beds often have four drawers underneath. Measure these carefully inside to make sure they are long, wide and deep enough to store the items you want to put there – a smaller drawer won't be big enough to take a spare double duvet or an average-sized suitcase, for example.

If you are planning a spare bedroom, or one for a child, you may want to have an extra bed for guests, or a friend to stay overnight, but for the room to normally be furnished with a single bed to allow maximum use of floor space. There are several beds around with a second stacking bed which slots below, and pulls out in a similar way to the drawers. Some types can then be raised on folding legs, or by a special mechanism to make them the same height as the original bed – others remain closer to the floor. One manufacturer specializing in a Victorian look in antique pine, makes a tall *chaise longue* with a bed-in-a-drawer which pulls out from underneath.

If you have an existing bed on legs, you may well be able to fit it with separate pull-out drawers on wheels or castors – or a folding bed to slip underneath. The result may not look very pretty, but you can fit the bed with a fabric valance (dust ruffle) to hide the drawers or bed from view. This may mean adding box pleats to the corners of the valance so that the legs are accommodated and the valance does not stick out awkwardly. Or you can simply make a valance which is not fully joined at the four corners to allow room for the legs.

The same trick can be used to hide suitcases, stacking crates, wire and other baskets and theatrical skips, or to conceal

transparent rigid plastic storage drawers and boxes, 'soft' plastic and even folding cardboard storage units. If you are buying any slot-under storage like this, remember to measure the height from the floor to the underframe of the bed – in other words the actual space you have to manoeuvre items in and out.

It may also be possible to enclose the base of such a bed to provide more permanent storage – it can be fitted with a frame, and louvred, panelled or flush doors can then be added; but the bed will not be easy to move when fitted with such a cumbersome base. In some instances an existing bed can have the legs removed and be raised on storage cupboards – or, in the case of a child's room, on toy boxes or a cabin-style frame.

Some beds may be high enough off the floor to accommodate some slip-under storage – look for narrow collapsible and instant storage boxes. Neat ready-made cardboard drawers which you cut and fold yourself; office-type box files; low baskets or cylindrical wicker containers for small items such as knitting wools and baby garments; or even bottles can sometimes fit under a slim bed base.

Above: Pull-out drawers form an integral part of a double divan base and can be used to store spare bedding and linen; winter clothes or sports and outdoor hobbies equipment. The drawers are easily reached without disturbing the made-up bed.

Right: *'Colour-coded' storage drawers in a child's room link with the scheme, and can be repainted when the scene changes. Use them to encourage tidiness – designate a drawer colour for specific items.*

Left: *Add a spare bed to a room without losing precious floor space. Here, a second single bed slips under the pine bedstead and can be easily pulled out to accommodate a guest.*

INSTANT STORAGE

Some of the built-in and permanent storage in this book may not always be practical for those on-the-move or who need more flexibility. And for those who need something immediately in which to store clothes, treasures, toys, a special collection, untidy papers, jewellery, scarves, belts or other accessories, 'instant' storage can often be an attractive answer to the problem.

There are many possibilities: clear plastic containers and 'sleeves' for sweaters, linen and shoes, which can hang behind cupboard doors, or slip under beds; brightly coloured simple plastic storage crates (some are actually designed to fold away when not in use, or to make transporting easier), which are cheerful enough to leave on show in a hi-tech scheme, children's rooms or modern kitchens; and amazing 'minimalist' pieces made from metal, wire and folded cardboard – some of which are available by mail order.

But however appealing and inexpensive a piece of instant storage is, you still need to follow the basic principles of planning or you may run the danger of adding more clutter to a room. Ask yourself whether it will fit into the space where you propose to put it. If you are going to stack crates, wire racks or storage drawers remember to measure height as well as depth and width. Also, make sure it will take the items you want to store comfortably – you do not want clothes to become too creased. So again, it means measuring up, doing a little arithmetic and taking the tape measure with you when you shop – or double-checking sizes in catalogues to make sure the item will serve its purpose adequately.

Left: Storage as an integral part of the room scheme: the basket provides storage for cushions, and ties in with the wicker furniture.

Below: A sturdy wooden chest provides traditional storage for bed linen and blankets.

STORAGE ON VIEW

In most rooms, instant storage becomes an integral part of the overall decorating theme – making a feature out of a necessity. Kitchen equipment can be suspended from special hooks, racks or chains and positioned close to the cooker or work surface; pulses, pastas and herbs can be stored in decorative transparent containers and soaps, bath salts and other toiletries can be similarly on show in the bathroom.

Wicker and other woven baskets in all shapes and sizes are usually decorative enough to be permanently on view in almost any room – woven willow cutlery baskets, covered with a pretty linen or checked gingham cloth can stand on a table, dresser or sideboard; a similar, or plain, basket can be used to store candles in the dining-room. Picnic hampers, usually neatly shaped like a suitcase, can be a practical place to keep table linens and cutlery in a dining area.

Above: *Make a guest room that bit more welcoming with an array of soap and toiletries in a flat wicker basket. The chest provides a dressing-table surface as well as storage space.*

Right: *Basically the simplest storage of all – a row of coat hooks, painted to contrast or blend in with the scheme. Large earthenware pots can be used in halls for umbrellas and walking sticks and in bathrooms for towels.*

Above: *Delicate wire baskets make a decorative addition to a country kitchen, and ensure utensils are close to hand. A traditional egg basket is used to store pastel-coloured kitchen cloths.*

Above: *A portable butcher's block-style trolley stores cooking implements. The wooden top can be used as a work surface in the kitchen and the trolley can be wheeled into the dining room, or out onto the patio for barbecues.*

Right: *Shaker-style boxes in primary colours come in sizes suitable for jewellery, accessories and even hats, and are attractive and covetable items in their own right.*

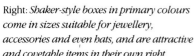

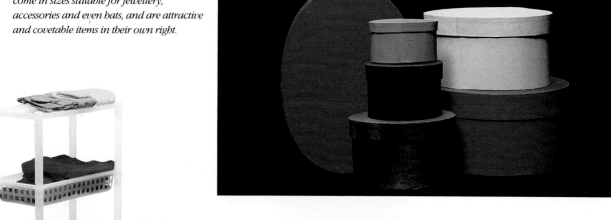

Left: *Organize bedroom closets with an array of storage ideas. Stackable shelving can be wheeled in and out of the cupboard, or used for bedside storage. Extra units can be added as required. Keep socks or other small items in sturdy plastic or cardboard drawers. The cardboard can be folded when not in use.*

Below: *Plastic stacking crates in cheerful colours might persuade even the most untidy child or teenager to put things away. They can be piled up to form a unit or bedside table and put out of sight under the bed.*

Above: *This plastic crate packs flat so it can be transported easily and stored when not needed. When opened up it makes a delightful toy box.*

Above: *This metal toy box is sturdy enough to withstand any amount of rough treatment, and the lightweight lid is perfect for small hands.*

Circular, oval, square – any interestingly-shaped baskets can be left natural, or painted with a matt paint in a pastel or subtle colour, to echo a bedroom or bathroom scheme, and used to hold soaps, cosmetics, belts, beads etc. An old bicycle basket with a curved front and flat back is ideally shaped for mounting on a wall. It could be used in a hall for keys, clothes brushes, family reminders and torches (flashlights); in the kitchen to store wooden spoons, whisks and spatulas; in the bedroom or bathroom for soaps, jewellery, cosmetics, or accessories or even used as a decorative container for flowers or plants.

Shopping, linen and log baskets (with handles) can be suspended from hooks, hung on a hat stand or lifted high up towards the ceiling on an old-fashioned rack-and-pulley, and used to store all manner of items. Special wire baskets, racks and sieves, suspended from a hook in the kitchen or dining-room can be filled with fruit and vegetables and look decorative – or used functionally to dry flowers and herbs for winter use.

Don't discount the humble hook – mounted behind a door; screwed into the front of a shelf or rack; or several, fixed to a neat piece of wood and mounted on a wall, can provide the simplest instant storage of all. Hang clothes, belts, scarves, jewellery, cups and mugs, bunches of dried flowers, or one of the shoe or sweater 'tidies' previously mentioned.

That old favourite – the bentwood hat stand – can also be a good buy for rooms where space is at a premium. Use it to hang and store scarves, belts and pieces of clothing, as well as hats, walking sticks and umbrellas.

Larger items with lift-up tops such as trunks, ottomans or square linen baskets double as coffee tables in the living room, as bedside tables, or dressing-tables.

Trolleys, either made of wood, metal, or with wire racks, can also make good instant storage, which is also portable, as they are usually fitted with wheels or castors and can be pushed from room to room. They can be used to house baskets, crates or wire racks to keep everything conveniently to hand.

Most of these instant storage ideas are easily portable and fairly long-lasting. However there are some storage boxes, chests and drawers available which are made of cardboard or even strong paper. These usually arrive flat, and you have to fold and fix them yourself. These come, in the main, from specialist shops and mail order suppliers, and are obviously less strong than the wood, wicker or wire equivalent, but make excellent temporary tidies.

SIMPLE STORAGE

Storage need not be complicated or specifically designed as hefty built-in pieces for particular areas. Some of the best storage ideas are the simplest; many are easy, instantaneous and inexpensive to install. Some good basic storage often 'just happens' because of a sudden flash of inspiration, and the need to use what is readily to hand.

Off-cuts from do-it-yourself decorating and carpentry jobs may be used in various ways. Small lengths of curtain pole can be mounted on a wall, or under a shelf, and used for hanging all manner of small, difficult-to-store items. Curtain track can be similarly used, with hooks threaded through the eyes in the runners. This can be combined with a fabric valance, or a small pelmet, if the track is ugly. Pieces of dowelling or beading can also be fixed across an alcove or recess and used in the same way.

Use a spare plank of wood by fixing wooden, brass or other decorative hooks or pegs to it at intervals. Fix it to a wall with a batten and use to suspend children's toys, shoe and sports bags, umbrellas and walking sticks, clothes, scarves or jewellery according to your needs and requirements. Cup hooks can provide the simplest and cheapest storage of all. Fix them underneath or to the edges of existing shelves, onto a wooden batten, or inside cupboard doors.

A family notice or bulletin board can be made from an offcut of hardboard or plywood, surfaced with spare cork tiles to take drawing pins (thumb tacks). Use a spare piece of pierced hardboard to form a pegboard to suspend small items such as keys on gaily coloured pegs. A softer piece of board (Cellotex or softboard) might be covered with felt or baize, held in place by decorative brass-headed upholstery tacks. Stretch a lattice of strong fabric tape diagonally across it, or front it with a piece of garden trellis to hold notes, letters, works of art, shopping lists and other small items. For safety's sake, always make sure such boards are firmly fixed to the wall, since the tendency is to overload them.

Basic items can also be adapted to provide storage space – an old wooden stepladder of the kind with treads on both sides can be painted in an attractive colour and used for display and storage – this can be expanded further by slotting wooden shelves (different lengths to suit the treads) across the ladder. Carpenters' 'hop-ups' can be similarly used; builders' trestles are another useful, and portable method of creating instant storage for heavier items, especially if combined with a firm top (an old flush door can be used in this situation).

STYLISH STORAGE

Storage forms an integral part of the furnishing and accessorizing of a room; certain decorating styles rely on the storage system (and items being stored) being on view to add character and interest to the scheme. A country-style kitchen for example, looks good with wicker storage baskets full of herbs and dried flowers, or cutlery and *batterie de cuisine*, protected with crisp gingham covers. Wooden spoons, cooking forks and whisks can be stuck in a large crock or flower pot. Glass jars full of pastas, pulses, dried fruit etc, add practical decoration to any type of kitchen.

Above: *An inventive use of a household object! The treads of the painted stepladder are extended with planks to form shelves for highly original and easy-to-adjust storage.*

Right: *A small bathroom crammed with good storage ideas – hanging baskets store spare towels and soaps and a small triangular table has a convenient lower shelf and holds pots, bottles and jars for bathroom needs.*

An old metal or wooden bedhead, a fireguard, fender, or other decorative recycled item might be suspended above a worktop or hung from the ceiling. Chromed steel bars, hooks or even a length of chain can be fixed in the same way, and all be used to suspend pots, pans, cooking implements etc in a modern setting. A wrought-iron frame can be similarly adapted. Some old kitchen and scullery favourites are making a comeback, such as ceiling airers on pulleys and plate draining racks. These can be used for decorative storage just as easily as for the original purpose – and in rooms other than the kitchen or utility room.

The hi-tech style makes a feature of industrial and basic shelving, wire racks, files, etc with all the stored items neatly on view, adding colour and textural interest to the scheme. Some modern interiors use industrial or domestic streamlined wheeled trolleys as portable and open storage for bottles, glasses or items for the dining table; and also for hi-fi systems and storing records and tapes; as well as for jewellery, cosmetics and accessories in the bedroom or bathroom.

SOFT STORAGE

Shoe bags, duffle bags, fabric rolls, a fabric panel of pop-in pockets – all make practical storage for small items. They are particularly

Left: *This free-standing build-it-yourself unit has been extended with side flaps to provide decorative extra storage and make a practical room divider.*

Below: *A metal bar converted from an old bed frame is hung from ropes suspended from the ceiling. Essential kitchen utensils are easily within reach above the work surface.*

Above: *Storage is provided in a nursery by small wire baskets and soft items hanging from hooks.*

good children's things, because they are completely safe. They just need hanging from a peg or hooks or by one of the methods previously described; or they can be suspended behind cupboard (and other) doors, as can larger 'wardrobe bags' for clothes. Soft storage bags in the nursery can take the shape of a circus tent or puppet theatre, and be used for small garments, nappies (diapers), toiletries and other junior essentials. In a toddler's room it might be decorated with cartoon characters.

Some of these items can be bought ready-made, but it is much more fun to make your own from remnants of furnishing or dressmaking fabrics, using various sewing techniques such as appliqué, patchwork, embroidery – even knitting or crochet if that is your forte (soft shoe bags can be knitted or crocheted for example). Personalize them with shapes and patterns (they might be appliquéed or embroidered with the owner's initials or name), or use fabric and colour to co-ordinate them with the room scheme.

SHAKER PEG RAIL

Left: Peg rails can extend as far along one wall as you wish, or could be used on all four sides of the room. Use pegs for hanging baskets, articles of clothing, decorations, or even – like the Shakers themselves – furniture.

▲ **1** Mark the centre positions for the pegs on the front face of the rail, on the centre line of the rail. If two rails will meet in a corner, mark the first peg position at a distance of at least 1½ times the peg length from the corner.

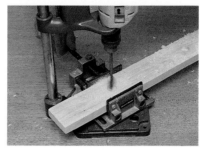

▲ **2** Drill blind holes 10mm (⅜in) deep for the pegs. Drill countersunk clearance holes for No.12 wall fixing screws through the centre of the two outermost holes and through the peg holes at 600mm (24in) intervals ·

▲ **3** Chamfer the edges and ends of the rail with a plane. Mark guide lines for the chamfer with a pencil, using your thumb as a gauge. If two rails will meet at a corner, cut those ends to a 45° mitre.

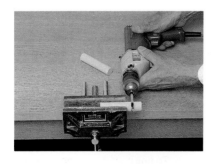

▲ **4** Cut the dowel pegs to length (100mm/4in approx) and chamfer the front end. Cut a small groove near one end of each peg to prevent things slipping off it. A drill file makes this task easy. Smooth with fine sandpaper wrapped around an offcut of dowel ·

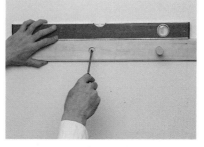

▲ **5** Drill the wall for wall fixings to take 50mm (2in) × No.12 screws, corresponding with the clearance holes in the rail. Insert the fixings and screw the rail to the wall. On a stud (dry) wall, the screws should go straight into the timber studs.

▲ **6** Apply PVA woodworking adhesive (white glue) to the inside of the peg holes and tap the pegs into place with a mallet. Check that the pegs are square to the surface and wipe off excess adhesive with a damp cloth.

BULLETIN BOARD

Left: *A bulletin board is useful in almost any part of the home. This one is used in a child's room, but you might want to hang one in the kitchen, study or home office; behind a door; or near the telephone.*

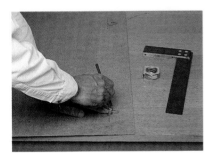

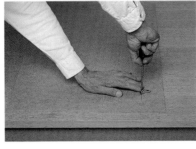

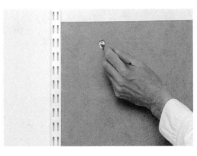

▲ **1** Cut the board to fit between the uprights. Mark the positions of the slotted mirror plates on the board. For each plate, drill two holes just larger than the head of the fixing screw right through the board and cut between them with a padsaw.

▲ **2** Screw each mirror plate to the board, slotted hole upwards, so that the top of the slot is just below the top of the slot in the MDF.

▲ **3** Position the board on the wall and mark hole positions for wall fixings to coincide with the top of the slots in the mirror plates. Drill the holes and insert the wall fixings. Insert roundhead screws until the head is slightly more than the thickness of the plate from the wall surface. Test the board for fit on the fixing screws, tighten and remove.

▲ **4** Lay out the tiles on the board, with the whole tiles in the middle, and adjust as necessary to give the widest possible border tiles. Mark the positions of the whole tiles. Spread cork tile adhesive over the front of the board and stick down the tiles, following the manufacturer's instructions.

▲ **5** Turn over the board and trim the edge tiles flush with the edges of the board, using a sharp trimming knife.

▲ **6** To finish off the top and bottom edges of the bulletin board, glue corner moulding such as birdsmouth to the board with PVA woodworking adhesive (white glue). Take care that none gets on to the surface of the tiles.

STORING THE WORKS

A hobby can call for special storage for all kinds of things – anything from bulky sports, gardening and gymnasium equipment, through knitting and sewing machines, or painting and drawing equipment to glues, small tools, tiny brushes and pots of enamel for delicate modelling. You may also have a home office, even if you don't work from home, in which case you will want to accommodate account books and files. Many homes now have a computer, which will require a deep desk, and shelving for its printer, disks and so on.

In an ideal world, you would probably aim to have custom-built storage to suit your particular work or hobby, sited in a separate area, and preferably with a lockable door; but all too often this is unrealistic, and the home office and hobbies facilities have to be fitted into another, already well-used area of the home. This is not as problematic as it might appear. With planning it is possible to accommodate everyone's requirements.

Left: *This storage wall is tailored to sewing, knitting and handicraft hobbies and also doubles as a home office; the cutting table/desk can be folded away to give additional floor space.*

SETTING UP

If you actually work from home, then it is sensible to try and set aside a separate room for this, so you can keep everything tidy, private and away from the rest of the family (small inquisitive fingers, pets etc). As well as privacy and adequate storage, you should consider the practical aspects of heating: it may be better to have some form of instant heating, rather than running the whole-house system all the time.

Also consider lighting. You will need an adjustable lamp for a desk, which does not cause glare or reflections in a VDU screen; lighting for inside cupboards, to see files; and controllable background lighting. If the nature of your work means that clients are invited home, try to have softer, more sympathetic lighting in the 'interviewing' area of the room.

The same comments apply to any hobbies areas and storage. The lighting will need to suit the purpose and function of the particular hobby, and be excellent for any close or detailed work. Aim to make good use of natural daylight, as well as installing adequate artificial lighting. This may involve some professional advice and installation. If you use a shed or garage as a work room, the electricity supply will have to be safely connected to it through buried heavy-duty cable. If in doubt, always check with a professional electrician.

You will also have to consider where to put the telephone (it must be handy for the desk), answering and fax machines and any other equipment. Sockets will have to be near at hand so you don't have any dangerous trailing wires or cables. If your hobby requires an electricity supply, for example for soldering or welding equipment, site the socket near the workbench. Also consider other safety points, such as blowtorches or caustic substances.

CHECK LIST

Once you have thought things through and decided where to site the office or hobbies area, you can then start to plan the space and storage facilities. You will need to measure up accurately and plan on paper.

If you are having to fit a home office into a living area or dining room, or spare bedroom, first consider any conflict of interest. Will a desk or computer be needed at the same moment as another member of the family is trying to watch television, practise a musical instrument, prepare the table and serve a meal, or go to sleep? This could make a nonsense of planning the space, and make it virtually impossible to create dual-purpose storage that works for everyone.

Similarly, it is no good having a hobbies workbench in the garage, which is to be worked at almost entirely in the evening, and when most probably the car will need to be in the dry, in the shared space. Hobbies facilities in a spare room, in the bathroom, on a landing, in an under-stair area or in a kitchen cupboard may equally cause conflict of interest and too much demand on the available space.

Once you have assessed the situation, you may find it makes sense to try and extend the property in some way – perhaps to enlarge and floor the attic or open up the cellar (in either case consider the ease of installing the flat-pack type of storage; *see* pages 92–93), or even to build on an extension, or have a separate building in the garden. The practicality of this will depend on the scale and shape of your house, the nature of the work or hobby, and of course the size of your budget. Both attics and basements or cellars are good places for hobbies such as model train layouts or other modelmaking, picture framing or winemaking; but not so sensible for a home office, which is in use all the time, or where good natural daylight is essential.

Nor are such confined spaces practical for hobbies which involve using heat or substances which give off fumes.

Attics can be roomy, and installing a sloping roof light, insulation and firm flooring may not be too costly, as long as the access is adequate. Think about the noise made by the hobby: musical instruments, active games, or the clatter of a typewriter might disturb those possibly trying to sleep below. A basement will probably lack natural daylight and may be damp and cold, but this could well be cured with some minor treatment and adequate heating. But don't use damp or very hot, dry areas to store items which will be adversely affected by heat.

Once you have decided on the most practical position for your office, workroom or hobbies area, make a list of all the things you will need to store. Think ahead and plan for the future – if your hobby is dressmaking, lacemaking, knitting or embroidery for example, you may well start off with modest equipment, but before you know where you are there will be masses of paper patterns, lots of cottons, threads, wools, embroidery silks and remnants of fabric to store. The office may start off simply, but computer

Below: A heavy-duty shelving bracket is used to create a hobbies corner. The sloping work surface can be folded back flat against the wall when not in use.

disks, tapes, files, folders, catalogues, reference and account books, office stationery – possibly samples – will soon all require storing.

Many hobbies call for special tools, as well as an adequate work surface. It makes sense to store tools as close to the work area as possible – above, to the side, or underneath. In some instances, tools and materials such as nails and screws can be stored in practical, portable containers under the bench, and they will be much easier to take on site when you are decorating, doing some carpentry or mending the motorbike.

Measure all these items you will need to store, so you can work out the most practical way of doing it, and the necessary size and shape of the storage units. In some cases it may be necessary to weigh some items (very heavy tools, for example), so you can be sure the shelves, drawers or racks will take them safely, without straining the bottoms of drawers or pulling shelf fixings out of the wall. As has been stressed throughout this book, it is no good building cupboards, having masses of drawer space or installing shelves if they don't work functionally, which means storing things without damaging them, and in such a way that they can be seen and reached easily and quickly.

Above: *A pull-out table is combined with tool storage above to make a neat workbench area and hobbies corner.*

Below: *A convertible table doubles as a desk; sliding doors hide shelves and home office items when not in use.*

HOME OFFICE

The starting point is usually a table or desk which you can use as a work surface for the typewriter, computer, drawing board or whatever, but also to provide some built-in storage. A conventional desk may well already incorporate drawers and cupboards – make sure these are as well organized as possible, with sectioned drawers, or pull-out deep ones with sliders for filing or partitions to separate one area from another. Again work out the size of the storage modules to suit your own requirements. Also remember that a desk for a computer needs to be at least 75cm (30in), preferably more, from front to back so that you can put the keyboard in front of the screen. If this is not feasible, you can buy or make a pull-out shelf for the keyboard – this must be firm and at a comfortable height.

Sometimes metal office desks can be bought cheaply from second-hand suppliers, as can some tatty old wooden ones. These can be given a new lease of life with spray paint for metal, or any suitable paint or stain for wood, as well as stencilled designs, new

doors or drawer fronts, or self-adhesive plastic chosen to suit your decorating scheme (*see* pages 55–58).

As an alternative to a desk, use two sets of drawers, or a set of drawers and a similar-sized cupboard, placed side by side with a gap between them for a 'knee-hole', and link them with a worktop. This can be a more flexible arrangement than a conventional desk, as it can be adjusted and the size of the top tailored to fit the space. The top can be changed later if you need to make a larger or smaller desk area, or to convert the arrangement to an L or U shape.

If you have enough space, you can use four drawers or cupboards, each set placed back to back, and with a giant-sized top, to create a 'partners' desk, where two people can sit each side, facing each other across the top. Check that the runners and bottoms of drawers in such an arrangement are strong enough to take the weight of office papers, which can be very heavy, and if necessary strengthen them before use.

If you are having to fit such an arrangement into a dual-purpose room, you may well be able to slot the desk into a recess to the side of a chimney breast (dining, kitchen or living room storage, or a bedroom dressing table or cupboards, may well be fitted into the other side) with wall-mounted shelves above, to take files, books and papers. Adjustable shelving will be more flexible and sensible, as it allows for change of height or depth of shelves as your requirements alter (*see* Focus on Shelving). Again, consider the weight of items to be stored as well as size and scale.

Sometimes you can devote a whole wall to such an arrangement, and if necessary conceal it behind folding doors, sliding screens, blinds or a curtain. Of course working facing the wall is not so congenial, nor so good for the eyes as working in front of a window. It may be possible to use a bay or bow window (perhaps in an upstairs bedroom) to site the desk, which could convert to a dressing table with portable mirror when necessary. You will need to have enough space to store the computer or typewriter when not in use, either inside the desk or in a nearby cupboard, ottoman or other place.

Building your office into a cupboard may be a more practical alternative – on the landing, under the stairs (*see* pages 62–63) or in some other convenient corner. If you do decide on

this option, make sure the lighting is adequate and built into the arrangement from the start. A recess or two return walls could provide a framework, or you could simply use strong uprights to form the carcass; doors could be folding, so the whole area opens out when you are at work. Alternatively use sliding doors, blinds or a curtain to close the area off, which will save space as well as money. If you do use doors, make maximum use of them, with hooks, racks or shelves on the backs to provide more storage.

It might not be possible, or even necessary to have a desk. A folding drawing table might be more practical, though remember you can only stand at these. It could hang on the back of the door or on a wall, and be pulled out and unfolded when necessary. Some other form of folding or flap-down table could equally be well used as a working surface. The 'cupboard' itself will then have to be kitted out with all the necessary racks, drawers and shelves for your needs, and still leave space to accommodate the folded table, trestles or whatever. Again accurate measurements, a calculator and some graph paper may be a necessary part of the planning exercise.

This type of arrangement can also be adapted to make a compact, separate hobbies area – the desk or drawing board becoming, or being replaced by, a workbench. Some folding tables and drawing boards are stout enough to take sewing or knitting machines

Above: *This is an alternative idea for a home office. Streamlined bedroom units are adapted to provide storage and desk space while bookshelves slot in under the window. The whole can be returned to bedroom status when necessary.*

or other equipment. If you need a really heavy-duty workbench, this will have to be incorporated into the design – it may be possible to have one which pulls out on wheels or castors, or on some form of sliding mechanism. An alternative is a folding portable bench.

Use the space underneath for storage. Stacking crates, or some form of pull-out or portable storage, may be the most sensible choice here, or a wheeled trolley which can hold items used at the bench. Again consider the safety angle of such an arrangement. When the doors are open, will it block access to the rest of the house? If the workbench is used for hot jobs, some form of fireproofing of the doors and inside the cupboard may be needed. Keep a fire extinguisher of the right type ready to hand.

GARDENING AND GAMES

If your hobby is an outdoor one, the conversion of a shed, greenhouse or lean-to, or the addition of a conservatory, may well provide the necessary space and storage facilities (don't forget to install an adequate electricity supply). Conservatories have been known to double up as billiard and snooker rooms, flower-arranging areas and utility rooms. Any built-in storage will need to be aesthetically pleasing, especially if the area is to be dual-purpose and leads into the kitchen, dining-room or main sitting area of the house, so custom-designed units may be the only answer.

Again, garden sheds are not always ideal places for garden tools and perishable products, because of damp and rust. Garages can present the same problems, unless they are an integral part of the house, or adequate heating and ventilation can be provided; but they can be ideal places for the bulkier tools, even installed on ceiling-mounted racks, or on a wide shelf above an 'up-and-over' garage door. Fully glazed conservatories can be very hot in summer, and intensely cold and damp in winter, unless they are adequately double-glazed, ventilated and heated.

It can be a problem to find a place for garden tools, especially bulky items such as lawnmowers and wheelbarrows, if you don't have a tool shed or greenhouse. Careful organization of the space behind the back door could provide the answer, with hanging wire racks and a pegboard for small tools, shelves or narrow cupboards for dry goods, a tall cupboard for larger items, industrial slot-together shelving providing a work bench and staging, and a place for more bulky equipment. This can also be a good place to store sports gear such as golf clubs, fishing tackle, tennis and badminton racquets; but don't forget that any guns and ammunition must be stored in separate safe, lockable cupboards – it is vital that these are not stored together.

Small items, such as garden twine, raffia, trowels, garden ties, secateurs, labels, gardening gloves and so on, can be effectively stored in fabric 'pockets' on a textile backing, which can be hung on the wall or back of a door. Natural hessian, green baize and felt look particularly attractive, and might be made to match a gardening apron and gloves. (This soft furnishing technique can also be used for many other small items, such as reels of cotton, knitting needles,

wools and embroidery silks.) An old tapering saucepan stand can be a good place to store flowerpots during the winter months – and in summer it can go out onto the patio, with some of the pots planted with trailing ivy, lobelia etc.

Small hand tools and other items of gardening equipment can be kept neatly in flowerpots, stacked sideways one on top of another, or you can use small cut lengths of plastic or ceramic piping, as used for plumbing and drainage. A wooden wine rack or cane circular stacking rack can be converted for this purpose, as can trugs and baskets hung from ceiling-mounted hooks.

If you have a tall utility room, large kitchen, sun lounge, 'working' conservatory, or space in the garage, this might be the place to fix a ceiling clothes airer hung on a pulley – these are ideal for suspending baskets filled with drying flowers and herbs, or items for flower arranging. They can also be a good way of storing folding garden chairs, beach and camping gear, tennis nets, croquet hoops and mallets, cricket stumps, skis and other

Below: A colourful way of storing garden tools – this outhouse has been adapted to provide practical and decorative outdoor storage. Paint matches the exterior brickwork.

seasonal games equipment, but don't overload them, and always make sure any overhead suspension is firmly fixed – and above all, safe.

It would not be practical to store big, bulky sports equipment which is used regularly on this type of rack, but sometimes portable dress rails, of the type seen in dress shops and on offer through mail-order catalogues, can be used to suspend games equipment – and they are particularly handy if they incorporate a shoe rack. There are also some industrial pull-out vertical racking systems which can be adapted to domestic use. Either of these rails or racks might be concealed in an under-stair cupboard, or fixed in a garage or porch, hidden behind a screen, tailored cover or curtain. Some similar horizontal shelving could be fixed above.

A large metal and wire mesh trolley, of the type used for shelf-stacking in supermarkets, or for baggage at stations and airports is also another good method of providing easily portable storage if there is room for one. They have quite hefty wheels, which is a great advantage in moving them. The trolley can be made more attractive with a tailored canvas or plastic cover to keep dust off the contents. Look in industrial supply and shopfitting catalogues; or ask your local supermarket about such items – you may be able to buy an old one direct, and give it a new lease of life with spray paint in an exciting colour.

Other shopfitting ideas can be equally well adapted to the home – screens with mesh racking, or pegboard and large metal hooks, onto which packets are slotted; grooved bases which take small bottles or other cosmetic products; racking for stationery. All could be adapted to tool, hobby and home office storage. Look in trade catalogues or ask your local shops for advice.

Above: A place for everything in a utility room which doubles as a hobbies area. Many items are wall-hung on special hooks to keep them neatly out of the way, or are constructed to fold away when not needed.

Below: *An architects' drawing table can be used in a domestic situation. The top folds down, and the table can be used for other activities.*

HANDY HOBBIES

If your hobbies require the storage of lots of different-sized items, from planks of timber and tools to tiny tacks, nails and screws, again vertical racking systems can be useful, as long as you have space for them alongside the workbench or in a convenient cupboard. But smaller things are often better stored in glass jars, or clear plastic containers so you can see the contents at a glance. Screw-top jars can be saved, then fixed to the underside of a shelf, so the contents are suspended, and can be seen at a glance, and reached in a moment.

There are also special storage boxes with small compartments, and various transparent stacking containers, on sale in hobbies outlets. Some items produced for tools, decorating materials and woodworking products can also be adapted for 'softer' hobbies such as handicrafts, patchwork, knitting, sewing and embroidery. So shop around for such adaptable items. Look in kitchen equipment departments and shops:

Above: *Carpentry and do-it-yourself tools are stored in a wall-mounted cupboard with sliding doors. The whole is slim and space-saving and could be mounted above a workbench or kitchen unit.*

Left: *Make the most of a well-lit area under the window – chests and units can be linked with a continuous laminated worktop to provide a desk, drawing and hobbies surface. The wall-mounted shelves to each side and above the window make the most of the available space.*

Left: A neat wooden box holds sewing 'notions', opening to form six separate compartmented drawers. The box closes for easy carrying by the integral handle.

Below: A fabric-lined basket will take work-in-hand and essentials and incorporates a cotton reel holder for easy organization.

many of the kitchen tool racks, plastic food containers, microwave bowls and dishes are ideal for use as alternative storage. A simple crock for wooden spoons and cooking forks, for example, will work equally well as storage for long knitting needles, crochet hooks, paintbrushes, upholstery tools, or a set of screwdrivers.

PORTABLE STORAGE

It is often easier to store items for knitting, sewing, tapestry, rugmaking or modelling in portable containers which can be taken with you from room to room, or into the car when you go away. A portable workbench is a useful item – this can be stored under a cover and brought out when needed, and it is an ideal place to stand a sewing or knitting machine. Put an old door on top and use this as a worktop if you need an extra-wide surface. Builders' folding trestles can be equally useful, and even a stout folding ironing board can sometimes be used depending on the activity planned.

Sewing boxes and baskets can be used not only for their intended purpose but also for small tools or pots of paint. You could equally well use a sectioned basket intended for bottles and glasses, or line an ordinary deep basket with fabric, with pockets in the lining to take small items. A fabric jewellery 'roll' and some cosmetic bags can be useful for small items of all kinds.

A board or piece of pegboard with long nails or lengths of stiff wire fixed into it can hold reels of cotton. This can be fitted with a handle, or placed in the bottom of a square basket leaving space for sewing needles, scissors, mending equipment at the side. Or you could make a soft fabric bag for these and tie it onto the handle.

You can use a metal portable tool box to take sewing and embroidery equipment, but it will be fairly heavy to lug around. Adapt a

linen box or basket, or old trunk or picnic hamper to easily portable storage by fitting it out with small lift-out grooved or moulded drawers. Leave space for other containers, and try to incorporate a sectioned drawer for reels of cotton.

You can use plastic stacking crates in a similar way, as long as you can organize the interior efficiently, some of the folding card, basketware and plastic storage units (*see* pages 72–75) can also be used.

TOOL RACK

Above: *The base board for a tool rack can be made of chipboard (particle board) with or without prefabricated holes. Everything will be immediately to hand if hung on the wall, rather than kept in a box or bag.*

▲ **1** A convenient size for the board is 900 × 600mm (36 × 24in): even with the tools in place it is still possible to carry it. Lay the board flat and arrange the tools on it to best advantage. Bear in mind that any tools supported on hooks will need a little clearance above to enable them to be lifted off. Similarly, those in spring clips need some clearance at each side for the clips to expand. When you are satisfied with the layout, mark the outline of each tool on the chipboard.

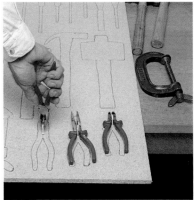

▲ **2** Decide on what type of support you will use for each tool (clips or hooks) and drill fine pilot holes for the screw threads. Square cup hooks can be used singly or in pairs, depending on the item to be supported. Plastic-coated ones are better for metal tools to avoid the danger of corrosion. Pad the jaws of pliers used to tighten them to prevent damage to the plastic coating. Screw each spring clip to the board in a position where it will grip the tool handle without allowing it to slip.

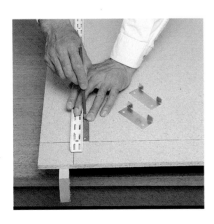

▲ **3** Screw the shelving uprights to the wall. Measure the spacing between the slots on the uprights and mark the back of the board accordingly. Lay a spare upright on the board with the slots centred on one marked line, and position the cabinet brackets in the slots. Mark the bracket and hole positions on the board. Square a line across to the other upright position and repeat the process.

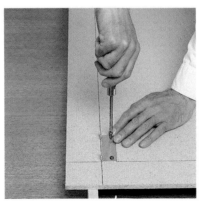

▲ **4** Screw the cabinet brackets to the chipboard with chipboard screws. Test the board on the shelving uprights to make sure that all the brackets engage fully and that it is properly supported. Hang the tools in position.

▲ **5** Cut the bottom shelf to length and screw shelf brackets to it, ensuring that they match the spacing of the uprights. The shelf is useful for storing bulky items and for holding tools in use off the board. It is not safe for this shelf simply to rest on the brackets – it must be screwed to them.

STORAGE LIGHTING

One important aspect when planning storage, and one which is often forgotten, is providing adequate lighting. It is no good having deep cupboards, elegant display shelves in alcoves, or clever fold-away systems if you can't actually see what you have got stored, or are unable to enjoy an eye-catching arrangement fully because it can't be seen at night.

This will mean doing some practical planning at the outset. You will need to make provision for wiring, fittings etc, and it may involve calling in an electrician. Think this through and discuss it with the carpenter, or built-in furniture specialist, if you are having the items custom-made – or make arrangements for lighting if you are installing your own storage. Some free-standing furniture comes with integral display lighting, but again this needs to be connected up to the power.

Above right: In this country kitchen, concealed spot lights illuminate display shelves. Strip lighting under wall-mounted cupboards light the work surface below.

Right: In a garage doubling as a workshop there is good natural daylight from glazed roof panels; night lighting is provided by a barrage of spots above the work bench and tool storage racks.

INSIDE STORY

If you have deep wardrobes or walk-in cupboards, a dark pantry or larder, a big enclosed storage area under the stairs, or any other large storage cupboard, it is essential to light it inside so you can use it efficiently. A simple lamp holder and bulb in the top of the cupboard is usually perfectly adequate, although larger runs of wardrobes might need two or three along the length, or a pair of lights one at each side of the area. Consider the heat given off by the bulb – sometimes a small fluorescent tube, which is much cooler, is a more practical option, especially if the light fitting has to be positioned close to clothes. It is also possible to install small cylindrical heaters, which should be put at the bottom of the cupboard, because hot air rises.

Often you need to open a cupboard with full hands, so if possible fit a switch to the door jamb or frame, so the light is automatically switched on as the door opens. If this is not practical, then fit the switch just outside the door – sometimes it can be positioned on the side of a projecting built-in. In a tall cupboard you can use a pull-cord switch on the light fitting.

Right: *Down lighters are fitted above wall-mounted cupboards to light the interiors and the work surfaces below; lamps on rise-and-fall fittings are mounted over the curved breakfast bar in this contemporary kitchen.*

Below: *Soft background lighting is provided in a traditional sitting room by lamps, wall up-lighters and display lighting in the glass-shelved cabinet; all are on separate circuits for flexible control.*

DISPLAY LIGHTING

Open storage shelves, recessed alcoves, glass-fronted display cabinets, and streamlined modern storage systems all look better if they have integral accent lighting. This should give as true a colour rendition as possible, so low-voltage fittings or tungsten-halogen lights may be advisable – avoid a cold, harsh light which may cause glare, and if you have to resort to fluorescent strip lighting, use a 'warm white' rather than the harsher cool 'daylight' type.

Again, if display shelves, cupboards or units are being built in, think about lighting at the planning stage. A recessed light in the top of an alcove, hidden behind an arched top, pelmet or coving will light shelves softly from above, but you will get some shadows on the lower shelves; glass, or other transparent shelves look magical illuminated from below, so a simple uplighter in the base of a unit may be better – but again consider the heat produced in any enclosed space.

In some cases, narrow strip lighting edging shelves is a better option – but this must not cause glare, so it may have to be concealed behind a baffle or batten. There are low-voltage flexi-strip systems which can be used to outline shelving.

Basic shelving systems installed in alcoves can be illuminated from the side, or above, by downlighters, wall-washers or spots as previously suggested; but it is also possible to buy individual spotlights which clamp onto shelves, creating a cosy glow on a group of items over a small area. This system of lighting can be effective if it suits the style of shelving and the room, and if you don't use too many – take care to avoid too many trailing flexes.

In the bathroom or bedroom the display and storage can be combined with lighting for a mirror. Some cabinets come with an illuminated mirror, but this is often inadequate. Try to arrange the lighting so it illuminates the face clearly – not the mirror – without glare, dazzle or shadow. Again shelves round the mirror with some form of built-in lighting might be the answer, or simple spotlights. Remember that lights must be enclosed in a bathroom, where steam or condensation can present a safety hazard, and that switches must be ceiling-mounted with pull cords.

If this is not practical, external lighting can be used to light storage. A row of ceiling-mounted downlighters positioned in the ceiling just above a run of cupboards can help to illuminate the inside; wall-washers, similarly mounted, can bathe the inside with an even light. Spotlights mounted on track above or to the side will also work reasonably well if they are angled correctly.

SAFETY PRECAUTIONS

There are now some EC directives concerning freestanding furniture which is supplied with built-in lighting fittings. Seek the advice of the supplier about this – it may seem unnecessary bureaucracy, but it is an extra safety precaution.

FLAT PACK STORAGE

Nowadays it is possible to buy complete shelving kits, kitchen units and cabinets, wardrobes, desks, chests, hi-fi racks and many other items in handy flat-pack form ready for home assembly. This is sometimes called KD, or 'knocked-down', furniture. It is usually displayed at the stockist or Do-it-Yourself superstore already made up, and with the instruction leaflet and several packs beside it, so you can assess how bulky a package you are going to have to carry home, and get some idea of how easy it will be to assemble.

Some companies will arrange home delivery (essential for a bulky item like a wardrobe); others will lend you a roof rack to put on to your car to help transport the item. It is wise to consider the strain this will put on your car roof, especially because most modern cars lack external gutters and have to use a rack with rubber feet, which can dent the roof. If necessary, arrange for a van to deliver the piece for you; but this will add a delivery charge to the cost of purchase.

As with all other storage, you will need to do some accurate measuring, both of the inside and of the outside of the piece, to make sure it will actually store what you want it to, and that it will fit into the space you have reserved for it. If the piece is not made up, check measurements on the literature; but these rarely give inside measurements, so either ask to have the package unwrapped, and try to assess the inside volume (you can work this out roughly by measuring the depth of the boards), or arrange to buy it on a sale-or-return basis, so that it can be changed, or money refunded, if you find it is not right when you get it home.

Above: *A basic, simple storage shelving unit with slotted angle and pre-cut shelves can be used in spare corners for many different purposes – paint it to contrast with the walls, so it stands out, or to tone so it fades into the background.*

PUTTING IT TOGETHER

One of the advantages of flat-pack furniture and storage is that it is fairly easy to get into a property, so that it is ideal for furnishing attics, or other spaces where access is restricted – it can be taken through a trap door, and erected in the room. However, it is not always easy to take out again, especially after it has been painted, stained or finished so that dismantling it will mar its appearance.

The stockist, and probably the leaflet on display, will tell you such furniture is 'simple to install; child's play; so easy a child of five can build it; all you need is a screwdriver' etc. But this is not necessarily true. First of all ask

to see the instructions, and check exactly what is needed in the form of screws, nails, glue and so on, and whether they are included in the pack. If not, make sure you buy the right quantity and size at the same time. There is nothing more maddening than getting a piece home and finding you can't assemble it immediately.

Also check what sort of screwdriver is required – quite often two or three sizes or types (usually including a Phillips or Pozidriv cross-headed one) are needed to put together one simple piece of furniture. And also check to see whether any other tools are necessary; if you already have these, fine, but if not,

remember the extra cost is all adding to the price of the piece. In some cases, if specialist tools are required, it may be possible to hire them at a fraction of the cost of purchase (good-quality saws for example, if you have to do a lot of cutting).

It is also a good idea to open the package in the shop, if you are taking it home yourself, and to check that everything is there that should be – sometimes essential screws are missing, or the instructions. Also check that these actually are the instructions for the piece you have bought – it is not unusual to find you have been given instructions for assembling something else. If you are having

Above: *A ubiquitous self-assembly base unit from a decorating superstore can be used for kitchen, utility room – or even bedroom storage.*

Right: *Variation on the shelving kit theme – the metal angle and shelves in a child's room are combined with plastic storage crates for extra space saving.*

the item delivered, open the package and check all this before signing the receipt.

Before you begin construction, clear an adequate space. If you are building it in a garage, or in the hall or landing, make sure you will be able to get it through the door into the room when it is finished. Put all the components out carefully on pieces of old brown paper or newspaper, or on a plastic sheet or dustsheet, and check against the instructions which piece is which, and if it is the correct way up. If necessary number the boards, or lightly mark them in chalk. Unpack screws, nails, hinges and catches, and count them to ensure there are enough. Then put them very carefully into a lidded container, so they don't get lost.

Put the storage together following the instructions to the letter. It is usual to build the carcass first, but this will depend on the item; sometimes doors or drawers are assembled first. It is sometimes practical to paint, stain or otherwise finish the piece in some way before you put it together, but if you do this you won't be able to change it if the pieces don't fit.

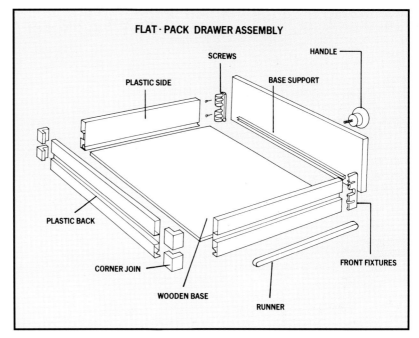

FLAT · PACK DRAWER ASSEMBLY

SCREWS
HANDLE
PLASTIC SIDE
BASE SUPPORT
PLASTIC BACK
CORNER JOIN
WOODEN BASE
RUNNER
FRONT FIXTURES

Above: *An 'exploded' diagram of a drawer shows how easy pack-flat furniture can be to build and install. Not all kits come with such graphic diagrams and you may find building-it-yourself taxes your ingenuity!*

STOCKISTS AND SUPPLIERS

UNITED KINGDOM

Black & Decker Ltd
Westpoint
The Grove
Slough
Berkshire SL1 1QQ
(power tools)

Cliffhanger Shelving
8 Fletchers Square
Temple Farm Industrial Estate
Southend-on-Sea
Essex SS2 5RN

Cubestore Ltd
58 Pembroke Road
London W8
(stackable storage systems)

Farbo-CP Ltd
Cramlington
Northumberland NE23 8AQ
(laminate for kitchen doors)

General Woodwork Supplies (SN) Ltd
76–80 Stoke Newington High Street
London N16
(timber, tools and hardware)

Hobbs & Co Ltd
88 Blackfriars Road
London SE1
(builders merchants)

Muji
26 Great Marlborough Street
London W1
(stackable plastic and cardboard drawers)

Spur Shelving Ltd
Otterspool Way
Watford
Hertfordshire WD2 8HT
0923 226071

Stanley Tools
Woodside
Sheffield S3 9PD
(woodworking, painting and decorating tools)

Strongbeam Co Ltd
65 East Barnet
New Barnet
Hertfordshire EN4 8RN
(shelving supports)

UNITED STATES

Allway Tools Inc
1255 Seabury Avenue
New York
New York 10462
(hand tools)

Bernhard Woodworking Ltd
3670 Woodhead Drive
Northbrook
Illinois 60062
(storage)

Black & Decker Ltd
Communications Department
702 East Joppa Road
Towson
Maryland 21286
(power tools)

Robert Bosch Power Tool Corporation
100 Bosch Boulevard
New Bern
North Carolina 28562
(power tools)

Home Building & Lumber Co Inc
1621 Avenue H
Rosenburg
Texas 77471
(lumber)

Shelves & Cabinets Unlimited
7880 Dunbrook Road
San Diego
California 92162
(storage)

CANADA

Seco Tools Canada
Mississanga
Ontario
Canada
(hand tools)

AUSTRALIA

Caringbah Sheet Metal (Aust) Pty Ltd
42 Cawarra Road
Carringbah
NSW 2229
(storage)

Parburys Building Products
30 Bando Road
Springvale
Victoria 3171
(wood merchants)

HJ Reece Pty Ltd
2 Barney Street
North Parramatta
NSW 2151
(hand tools)

Stanley Works Pty Ltd
PO Box 10
400 Whitehorse Road
Nunawading
Victoria 3131
(woodworking, painting and decorating tools)

NEW ZEALAND

Nees Hardware Ltd
11–15 Pretoria Street
Lower Hutt
(tools)

Stanley Tools (NZ) Ltd
PO Box 12–582
Penrose
Auckland
(woodworking, painting and decorating tools)

SOUTH AFRICA

Algoran Shelvit (Pty) Ltd
Botha Street
Alrode
Alberton
South Africa
(storage)

Tuf Tools CC Box 212
Maraisburg 1700
Transvaal
(woodworking, painting and decorating tools)

INDEX

PICTURE CREDITS

p6, Cuprinol/Leedex Public Relations; p7 top, Concord Lighting; p7 bottom, Dulux paints; p8, Alno UK Ltd; p9 top, Christies Fitted Bedrooms; p9 bottom, Arena of Oxford; p10, Ikea; p14, Camera Press; p15, Camera Press; p17 Camera Press; p20, Derwent Upholstery; p21 top, Crown Paints; p21 bottom left, Janet Mason/ Hayloft Woodcraft Ltd; p23 top right, Wickes Builder's Merchants; p23 bottom left, Sharps Bedrooms; p24, Alno UK Ltd; p25 top left, Arena of Oxford; p26, Sharps Bedrooms; p27 Wickes Builder's Merchants; p29, Janet Mason/Hayloft Woodcraft Ltd; p30, Camera Press; p31, Ikea; p32, Concord Lighting; p35, Richard Burbridge; p37 bottom, Elizabeth Whiting Associates; p38, Crown Paints; p41 top, Spur Shelving Ltd; p42, Spur Shelving Ltd; p43, Concord Lighting; p45, Strongbeam Shelving Co Ltd; p46 top, Elizabeth Whiting Associates; p46 bottom, Camera Press; p49, Camera Press; p50, Janet Mason/Hayloft Woodcraft Ltd; p51, Crown Paints; p53, Barkers Public Relations; p54, Infopress/Elrose Products; p55, Barkers Public Relations; p61, Crucial Trading Ltd; p62, Beaver and Tapley; p63, Camera Press; p64, Camera Press; p69 Beaver and Tapley; pp76–89, Camera Press; p90 bottom, Concord Lighting; p91 top, Grapevine Public Relations/Poppenpohl; p91 bottom, Condor Public Relations; p92, Camron Public Relations/Spur Shelving Ltd; p93 left, Wickes Builder's Merchants; p93 right, Camron Public Relations/Spur Shelving Ltd.

ACKNOWLEDGMENTS

The author and publishers would like to thank the following companies for the loan of materials used in this book. See Stockists and Suppliers for addresses.

Cliffhanger Shelving; Cubestore; Muji; Strongbeam Co. Ltd